Cement-Based Materials for Nuclear Waste Storage

Florence Bart • Céline Cau-dit-Coumes
Fabien Frizon • Sylvie Lorente
Editors

Cement-Based Materials for Nuclear Waste Storage

Springer

Editors
Florence Bart
Atomic Energy and Alternative
Energies Commission, CEA Marcoule
Bagnols-sur-Cèze, France

Céline Cau-dit-Coumes
Atomic Energy and Alternative
Energies Commission, CEA Marcoule
Bagnols-sur-Cèze, France

Fabien Frizon
Atomic Energy and Alternative
Energies Commission, CEA Marcoule
Bagnols-sur-Cèze, France

Sylvie Lorente
Université de Toulouse, UPS, INSA
Laboratoire Matériaux et Durabilité
des Constructions, Toulouse, France

ISBN 978-1-4614-3444-3 ISBN 978-1-4614-3445-0 (eBook)
DOI 10.1007/978-1-4614-3445-0
Springer New York Heidelberg Dordrecht London

Library of Congress Control Number: 2012941742

Printed on acid-free paper

Springer is part of Springer Science+Business Media (www.springer.com)

Preface

Operations in the nuclear power industry produce a wide variety of radioactive wastes. The predisposal waste management is to produce waste packages suitable for storage, transportation and final disposal. Cementation is a widely applied technique for the conditioning of low- and intermediate-level radioactive wastes. Compared with other solidification techniques, the cementation process is relatively simple and inexpensive. Moreover, calcium silicate cements show many advantages, such as easy supply, compatibility with aqueous wastes, good self-shielding and high alkalinity, which allow to precipitate and thus confine many radionuclides. However, specific issues have still to be addressed, such as the limitation of adverse cement–waste interactions which may affect the quality of the resulting solidified waste form, or the understanding and prediction of long-term properties of the waste packages and cement barriers. There is also a need for safe processes minimizing the production of secondary wastes.

Facing these major challenges, the objective of the 1st International Symposium on Cement-Based Materials for Nuclear Wastes (NUWCEM), that was held in Palais des Papes, Avignon and Marcoule Research Center on 11–14 October, 2011, was to promote the exchange of advanced information on the ongoing research and development activities dealing with cementation of nuclear wastes, from elaboration of waste packages to their final disposal and long-term behaviour.

This book contains some of the contributions to the NUWCEM Symposium. The invited speakers and the contributions from the participants covered the full range of theoretical, computational, experimental and technological approaches. In addition to the present contribution, there were numerous other excellent contributions to the meeting, but because of the shortage of space, not all of these presentations could be included in this book.

The chapters included in this book have been grouped as follows: methods of production of cement–waste forms, physico-chemical processes occurring in cement–waste forms at early age, influence of external and internal factors on long-term properties of cement–waste package and cement barriers and emerging and alternative cementitious systems.

In the first group of contributions, concerning methods of production of cement–waste forms, Drace and Ojovan summarize IAEA research project on cementitious materials for radwaste management; Sukhanov et al. discuss the main results of cementation technology for immobilization of hydroxide pulps and Cardinal et al. evaluate the performance of high energy mixer for the cementation of ILW surrogate slurry.

The works on physico-chemical processes occurring in cement–waste forms at early age compose the second group of contributions. Lothenbach investigates the blending of Portland cement with silica-rich materials that leads to changes in composition of the hydrated cement and of its pore solution; Gibsy et al. describe the application of a sub-lattice model for C-S-H gel, allowing substitution of alumina, sulphate or heavy metals into the structure, while Aimoz et al. develop a thermodynamic model for the AFm-(I_2, SO_4) solid solution with the purpose of predicting the fate of ^{129}I in the complex cement matrix.

In the third part, contributions concerning the influence of external and internal factors on long-term properties of cement–waste package and cement barriers are regrouped. Torrenti et al. examinate the coupling between calcium leaching and mechanical behaviour of concrete, while Bary et al. develop a chemo-transport-mechanical model for the simulation of external sulphate attack in cementitious materials. Concerning radionuclides sorption and transport, Wieland et al. present a study on U(VI) and Np(V, VI) binding in hardened cement paste by micro-X-ray fluorescence and micro-X-ray absorption spectroscopy (μ-XRF/XAS); Wang et al. summarize a significant database of selected sorption values on cement for 25 elements; Mercado et al. propose an experimental way to obtain the diffusion coefficient of ions based on the determination of the electrical conductivity of the specimen; Pang et al. review the multiphase modelling of moisture transport process in pore structure of cement-based materials, while Boher et al. focus on the hydrogen diffusion through partially water saturated materials. Finally concerning the concrete cells of nuclear repositories, Albrecht et al. review the current status of microbiological catalysis of redox reactions in concrete and Depierre et al. study the leaching of nuclear glasses in cement pore waters.

The final part will consider the emerging and alternative cementitious systems for radwaste solidification and stabilization. After an exhaustive review of such binders, proposed by Cau, a focus will be made on several matrices. Calcium sulfo-aluminate cements are used as $ZnCl_2$ stabilization matrix by Berger et al. and as highly concentrated borate solutions conditioning material by Champenois et al. On the other systems, Swift et al. evaluate the effects of supplementary pulverised fuel ash on phosphate-modified calcium aluminate cement, while Poulesquen et al. present the rheological behaviour of geopolymers. Finally on low-pH cement, Bach et al. propose an original approach to predict their long-term chemical evolution and Alonso et al. summarize the results of the international project Round Robin test, which aims to propose methodologies for pH determination in low-pH cement systems used in the context of nuclear waste repositories.

The Editors would sincerely like to acknowledge the financial support given by Areva, Ondraf, Nuclear Filter Technology, Endel GDF Suez and A3i.

We also would like to thank the members of the Waste Treatment and Conditioning Research Department of the Atomic Energy and Alternative Energy Commission (CEA) for all the hard work that many researchers have put into this venture. Finally, we would like to thank Pauline Benoit and Anne Ponge for the formatting of this book and the preparation of the figures. We are deeply indebted to them for all their care and attention, and the patience that they have shown in both the NUWCEM conference and the book preparation.

Bagnols-sur-Cèze, France — Florence Bart
Bagnols-sur-Cèze, France — Céline Cau-dit-Coumes
Bagnols-sur-Cèze, France — Fabien Frizon
Toulouse, France — Sylvie Lorente

Contents

Contributors

L. Aimoz Laboratory for Waste Management (LES), Paul Scherrer Institute (PSI), Villigen, Switzerland

Institute of Geological Sciences, University of Bern, Bern, Switzerland

A. Albrecht Andra, Châtenay-Malabry, France

M.C. Alonso Research Centre in Safety and Durability of structures and materials, CISDEM (CSIC-UPM), Madrid, Spain

F. Angeli Commissariat à l'Energie Atomique et aux Energies Alternatives, DEN/MAR/DTCD/SECM, Bagnols-sur-Cèze, France

G. Aouad Ecole des Mines de Douai, MPE-GCE, Douai, France

Université Lille Nord de France, Lille, France

T. Bach Commissariat á l'Energie Atomique et aux Energies Alternatives, DEN/MAR/DTCD/SPDE, Bagnols-sur-Cèze, France

ICB, UMR 5209 CNRS Université de Bourgogne, Dijon, France

F. Bart Commissariat à l'Energie Atomique et aux Energies Alternatives, DEN/MAR/DTCD/SPDE, Bagnols-sur-Cèze, France

B. Bary Commissariat á l'Energie Atomique et aux Energies Alternatives, DEN/DANS/DPC/SCCME, Gif-sur-Yvette, France

F. Benboudjema LMT/ENS Cachan/CNRS UMR8535/UPMC/PRES UniverSud Paris, Cachan, France

S. Berger Commissariat à l'Energie Atomique et aux Energies Alternatives, DEN/MAR/DTCD/SPDE, Bagnols-sur-Cèze, France

J.A. Berry Serco TCS, Didcot, UK

A. Bertron LMDC-INSA, Laboratoire Matériaux et Durabilité des Constructions, Université de Toulouse, UPS, INSA, Toulouse Cedex 04, France

C. Boher UPS, INSA, LMDC (Laboratoire Matériaux et Durabilité des Constructions), Université de Toulouse, Toulouse Cedex 04, France

Commissariat à l'Energie Atomique et aux Energies Alternatives, DEN/MAR/DTCD/SPDE, Bagnols-sur-Cèze, France

X. Bourbon ANDRA, Chatenay-Malabry, France

G. Cardinal Commissariat à l'Energie Atomique et aux Energies Alternatives, DEN/MAR/DTEC/SDTC, Bagnols-sur-Cèze, France

C. Cau-dit-Coumes Commissariat à l'Energie Atomique et aux Energies Alternatives, DEN/MAR/DTCD/SPDE, Bagnols-sur-Cèze, France

J.B. Champenois Commissariat à l'Energie Atomique et aux Energies Alternatives, DEN/MAR/DTCD/SPDE, Bagnols-sur-Cèze, France

N.C. Collier Immobilisation Science and Technology Team, National Nuclear Laboratory, Warrington, UK

M.A. Cuñado Empresa Nacional de Residuos Radioactivos, ENRESA, Madrid, Spain

E. Curti Laboratory for Waste Management (LES), Paul Scherrer Institute (PSI), Villigen, Switzerland

R. Dähn Laboratory for Waste Management, Paul Scherrer Institute, Villigen, Switzerland

D. Damidot Ecole des Mines de Douai, MPE-GCE, Douai, France
Université Lille Nord de France, Lille, France

R.H. Davies National Physical Laboratory, Teddington, UK

P. De Cannière Institute for Environment, Health, and Safety, Belgian Nuclear Research Centre (SCK•CEN), Mol, Belgium

T. de Larrard LMT/ENS Cachan/CNRS UMR8535/UPMC/PRES UniverSud Paris, Cachan, France

S. Depierre Commissariat à l'Energie Atomique et aux Energies Alternatives, DEN/MAR/DTCD/SECM, Bagnols-sur-Cèze, France

E. Deville Commissariat à l'Energie Atomique et aux Energies Alternatives, DEN/DANS/DM2S/SFME, Gif-sur-Yvette, France

A.T. Dinsdale National Physical Laboratory, Teddington, UK

Louis Donnet Commissariat à l'Energie Atomique et aux Energies Alternatives, DEN/MAR/DTEC/SDTC, Bagnols-sur-Cèze, France

Z. Drace International Atomic Energy Agency, Vienna, Austria

F. Frizon Commissariat à l'Energie Atomique et aux Energies Alternatives, DEN/MAR/DTCD/SPDE, Bagnols-sur-Cèze, France

Laboratoire d'Etude de l'Enrobage des Déchets, French Atomic Energy and Alternative Energies Commission, Bagnols-sur-Cèze Cedex, France

CEA, DEN, L2ED, Bagnols-sur-Cèze, France

X. Gaona Karlsruhe Institute of Technology, Institut für Nukleare Entsorgung, Karlsruhe, Germany

J.L. García Calvo Research Centre in Safety and Durability of structures and materials, CISDEM (CSIC-UPM), Madrid, Spain

S. Gin Commissariat à l'Energie Atomique et aux Energies Alternatives, DEN/MAR/DTCD/SECM, Bagnols-sur-Cèze, France

J.A. Gisby National Physical Laboratory, Teddington, UK

F.P. Glasser Department of Chemistry, University of Aberdeen, Old Aberdeen, UK

M.A. Glaus Laboratory for Waste Management (LES), Paul Scherrer Institute (PSI), Villigen, Switzerland

D. Jacques Institute for Environment, Health, and Safety, Belgian Nuclear Research Centre (SCK•CEN), Mol, Belgium

H. Kinoshita Immobilisation Science Laboratory, Department of Materials Science and Engineering, The University of Sheffield, Sheffield, UK

V.A. Kravchenko The Mining Chemical Combine, Krasnoyarsk Region, Russia

D.A. Kulik Laboratory for Waste Management (LES), Paul Scherrer Institute (PSI), Villigen, Switzerland

D. Lambertin Commissariat à l'Energie Atomique et aux Energies Alternatives, DEN/MAR/DTCD/SPDE, Bagnols-sur-Cèze, France

P. Le Bescop Commissariat à l'Energie Atomique et aux Energies Alternatives, DEN/DANS/DPC/SCCME, Gif-sur-Yvette, France

A.V. Lebedeva A.A. Bochvar High-Technology Research Institute of Inorganic Materials, Moscow, Russia

B. Leterme Institute for Environment, Health, and Safety, Belgian Nuclear Research Centre (SCK•CEN), Mol, Belgium

N. Leterrier Commissariat à l'Energie Atomique et aux Energies Alternatives, DEN/DANS/DM2S/SFME, Gif-sur-Yvette, France

C.Q. Li Key Laboratory of Safety and Durability of Civil Engineering of China Education Ministry, Tsinghua University, Beijing, China

Civil Engineering Department, Tsinghua University, Beijing, China

K.F. Li Key Laboratory of Safety and Durability of Civil Engineering of China Education Ministry, Tsinghua University, Beijing, China

Civil Engineering Department, Tsinghua University, Beijing, China

M. Libert Commissariat à l'Energie Atomique et aux Energies Alternatives, DEN/CAD/DTN/SMTM, Saint-Paul-lez-Durance, France

S. Lorente UPS, INSA, LMDC (Laboratoire Matériaux et Durabilité des Constructions), Université de Toulouse, Toulouse Cedex 04, France

B. Lothenbach Empa, Lab. Concrete & Construction Chemistry, Dübendorf, Switzerland

Swiss Federal Laboratories for Materials Science and Technology (EMPA), Concrete Construction Chemistry Laboratory, Dübendorf, Switzerland

N. Macé Laboratory for Waste Management, Paul Scherrer Institute, Villigen, Switzerland

D. Mallants Institute for Environment, Health, and Safety, Belgian Nuclear Research Centre (SCK•CEN), Mol, Belgium

E. Martens Institute for Environment, Health, and Safety, Belgian Nuclear Research Centre (SCK•CEN), Mol, Belgium

V.I. Matselya The Mining Chemical Combine, Krasnoyarsk Region, Russia

H. Mercado UPS, INSA, LMDC (Laboratoire Matériaux et Durabilité des Constructions), Université de Toulouse, Toulouse Cedex 04, France

C. Mercier LMCPA, Université de Valenciennes et du Hainaut Cambrésis, Valenciennes cedex, France

M. Naito Japan Atomic Energy Agency in Japan, JAEA, Aomori, Japan

N.A. Naumenko A.A. Bochvar High-Technology Research Institute of Inorganic Materials, Moscow, Russia

A. Nonat ICB, UMR 5209 CNRS Université de Bourgogne, Dijon, France

M. Ochs BMG Engineering Ltd., Schlieren, Switzerland

M.I. Ojovan Immobilisation Science Laboratory, Department of Materials Science and Engineering, University of Sheffield, Sheffield, UK

X.Y. Pang Key Laboratory of Safety and Durability of Civil Engineering of China Education Ministry, Tsinghua University, Beijing, China

Civil Engineering Department, Tsinghua University, Beijing, China

S. Pettersson Swedish Nuclear fuel and waste management, SKB, Stockholm, Sweden

I. Pochard ICB, UMR 5209 CNRS Université de Bourgogne, Dijon, France

O.N. Pogorelko A.A. Bochvar High-Technology Research Institute of Inorganic Materials, Moscow, Russia

A. Poulesquen Commissariat à l'Energie Atomique et aux Energies Alternatives, DEN/MAR/DTCD/SPDE, Bagnols-sur-Cèze, France

I. Puigdomenech Swedish Nuclear fuel and waste management, SKB, Stockholm, Sweden

P. Regnard Commissariat à l'Energie Atomique et aux Energies Alternatives, DEN/MAR/DTEC/SDTC, Bagnols-sur-Cèze, France

Yu. A. Revenko The Mining Chemical Combine, Krasnoyarsk Region, Russia

L.P. Sukhanov A.A. Bochvar High-Technology Research Institute of Inorganic Materials, Moscow, Russia

P. Swift Immobilisation Science Laboratory, Department of Materials Science and Engineering, The University of Sheffield, Sheffield, UK

Y. Takeshi Central Research Institute of Electric Power in Japan, CREIPI, Tokyo, Japan

J. Tits Laboratory for Waste Management, Paul Scherrer Institute, Villigen, Switzerland

J.M. Torrenti Département Matériaux, Université Paris Est, IFSTTAR, Paris, France

E. Tronche Commissariat à l'Energie Atomique et aux Energies Alternatives, DEN/MAR/DTEC/SDTC, Bagnols-sur-Cèze, France

M. Tyrer Mineral Industry Research Organisation, Birmingham, UK

H. Ueda Nuclear Waste Management Organization of Japan, NUMO, Tokyo, Japan

L. Vielle-Petit BMG Engineering Ltd., Schlieren, Switzerland

M. Vuorio Nuclear waste Management in Finland, Posiva, Eurajoki, Finland

C. Walker Japan Atomic Energy Agency in Japan, JAEA, Aomori, Japan

C.S. Walker Japan Atomic Energy Agency (JAEA), Ibaraki, Japan

L. Wang Institute for Environment, Health, and Safety, Belgian Nuclear Research Centre (SCK•CEN), Mol, Belgium

H. Weber National Cooperative for the Disposal of Radioactive Waste in Switzerland, NAGRA, Wettingen, Switzerland

E. Wieland Laboratory for Waste Management (LES), Paul Scherrer Institute (PSI), Villigen, Switzerland

K.P. Zakharova A.A. Bochvar High-Technology Research Institute of Inorganic Materials, Moscow, Russia

Part I
Methods of Production of Cement-Waste Forms

Chapter 1
A Summary of IAEA Coordinated Research Project on Cementitious Materials for Radioactive Waste Management

Z. Drace and M.I. Ojovan

1 Introduction

The IAEA Coordinated Research Project (CRP) on cementitious materials for radioactive waste management has started in 2007 and has gathered 26 research organizations from 22 member states to share their research and practices on the use of cementitious materials [1, 2]. The CRP facilitated the exchange of information and research co-operation in resolving similar problems between different institutions and contributed towards improving waste management practices, their efficiency and general enhancement of safety.

The objective of the IAEA CRP was to investigate the behaviour and performance of cementitious materials used in radioactive waste management system. These included waste packages, wasteforms and backfills as well as investigation of interactions and interdependencies of these individual elements during long-term storage and disposal. Fundamental to this is an understanding of the processes that lead to the degradation of the barriers, physical and chemical properties. The purposes of these elements in a waste management system are different, and thus the requirements for long-term performance and account of degradation mechanisms/consequences may also be different. However, all elements are important, as well as interactions envisaged between them, ultimately aiming to ensure the overall safety of a storage/disposal system.

Cement has many favourable properties, both chemical and physical, making it a desirable barrier or matrix for the encapsulation of radioactive and toxic wastes.

Z. Drace
International Atomic Energy Agency, Wagramer Strasse 5, PO Box 100, 1400, Vienna, Austria

M.I. Ojovan (✉)
Immobilisation Science Laboratory, Department of Materials Science and Engineering, University of Sheffield, Sheffield, UK
e-mail: M.Ojovan@sheffield.ac.uk

F. Bart et al. (eds.), *Cement-Based Materials for Nuclear Waste Storage*,
DOI 10.1007/978-1-4614-3445-0_1,

Chemically, cement has a high pH (e.g. its pore water) and forms hydration products which favour sorption and ion substitution. Physically, cement is a durable solid material with a low permeability in its hardened state which protects the radioactive waste, facilitating its safe transportation and storage. Cement is also an inexpensive and readily available material, durable in its hardened state, fluid when initially cast and tolerant to a variety of waste forms, including those in solid and liquid states. Cements have also been proved to show their stability when irradiated and ability to act as radiation shielding. Conventional cementitious materials such as ordinary Portland cement (OPC) and OPC-based composite cements, as well as novel cement systems such as geopolymers, high alumina and calcium sulfoaluminate and MgO-based or phosphate (acidic) cements can be used to create reliable immobilizing elements for safe storage and disposal of wastes. These barriers function, as well as interactions envisaged between various components, were considered in this CRP which was focused on predisposal management systems. The main research outcomes of the IAEA CRP are summarized under four topical sections: (a) conventional cementitious systems; (b) novel cementitious materials and technologies; (c) testing and waste acceptance criteria; and (d) modelling long-term behaviour, as well as conclusions of the overall CRP.

2 Conventional Cements

Conventional systems for immobilization of radioactive waste consist of cement, waste and water [3]. The order of mixing is often important but in some cases is limited by the nature of the waste, e.g. infiltration of scrap by a fluid cement—water mix. Liquid and particulates are often homogenized with cement by stirring. In many applications, a portion of the cement is substituted by a supplementary cementing material, e.g. fly ash or slag. After setting, the conditioned product can be transported and disposed of in a facility. Apart from the cementitious waste products (CWPs), cement is also used for construction of engineering barriers such as steel-reinforced trenches, silos, etc. These, together with aspects of steel corrosion, have been addressed in this CRP.

The cement powder is itself not homogeneous. The industry varies the proportions of these phases to influence the strength gain and increase resistance of the hardened product in subsequent service to attack by soluble sulphate. Cements are most often used in engineering applications as composite materials. The components of a CWP are complex. Thus, the main variables in optimization of cement matrix are (1) waste, (2) water-to-cement ratio, (3) waste-to-cement ratio, (4) admixtures (type, content), (5) admixture-to-cement ratio, (6) type of cement, (7) order of mixing and (8) emplacement and curing. In process optimization, it is usual to change one variable at a time. The formulation is made in a suitable container with the requisite quantity of waste, cement and water (if required). The order of addition is often important to ensure a homogenous product. Using liquid and particulate wastes, the mixture is homogenized most often (but not only)

Table 1.1 Example of conventional cementitious formulations

Waste stream	Cementitious matrix[a]
Spent ion exchange resins	Slag-Portland blends
Sludge and concentrates generated from treatment of LLW, incinerator ash	OPC, with or without additives
Mixture of sludge and ion exchange resin	OPC, Slag-Portland cement
Intermediate level liquid waste	Slag-Portland blends, OPC with vermiculite
Secondary waste generated during treatment of spent solvent from reprocessing plant	OPC
Waste generated during reprocessing of thorium-based spent fuel	OPC
Evaporator concentrate containing boric acid	OPC

[a]"Admixture" includes supplementary cementing materials such as slag or fly ash

using a mechanical stirrer. The stirrer is removed and the CWP is allowed to set undisturbed for a few days. Thereafter, the CWP may be removed from the container and curing continued in a humidity chamber at ~100 % RH or under water. After complete curing, nominally achieved in 28–90 days, the monolith can be used for leaching and compressive strength measurements.

Cement matrices have been used for immobilization of various types of wastes (Table 1.1). These CWPs are primarily evaluated for compressive strength and chemical durability. Other characterization procedures include (1) phase identification by XRD, (2) porosity, (3) water resistance, (4) freeze–thaw resistance, (5) flowability, (6) heat generation and (7) loading of salts.

To verify the durability of the waste forms, several leaching methods have been defined but none have been yet subject to sufficient critical scrutiny to be considered as standard. The conclusion of the participants to CRP is to promote standardization of durability verification. Tests are available for both static and dynamic conditions, but irrespective of the test method, analytical work is required following leach tests. Variables affecting the leaching rate during testing have to be agreed as appropriate. Defining "appropriateness" creates problems where data have to be kept generic because no site has been defined and the expected hydrogeological conditions are not known. Therefore, the main variables to be considered are the flow rate, the time of leaching, the effects of the temperature and the leachant composition.

3 Novel Systems

The CRP has discussed progress achieved so far on four types of alternative cementitious systems as shown in Table 1.2.

The most engineering experience has been achieved with CAC and C$A cements. The geopolymer type materials such as SIAL are made from alkali silicate (sodium, potassium) and metakaolin which is produced from kaolin by calcination

Table 1.2 Representative non-Portland cement systems

Designation	Formation conditions	Comments
SIAL	Mixture of sodium silicate (hydrate) with metakaolin	Geopolymer-type matrix which is characteristically X-ray amorphous
Magnesium phosphate cement	Mixture of fine grained MgO (periclase) and a phosphate source, e.g. phosphoric acid or monopotassium phosphate (MKP)	Many variants are known, differing in pH and solubility. Not fully commercial except for small-scale applications, e.g. as refractory or dental cements
Calcium sulfoaluminate cement (C$A, where $ = SO_3)	Available commercially or made by mixing commercial calcium aluminate cement (CAC) with calcium sulphate	Has a history of use (~40 years) as a construction cement. Developed in China but now widely available
Calcium aluminate cement (CAC)	Based on clinkers or fused products with dicalcium silicate and $CaAl_2O_4$	Calcium aluminate cements are widely available as commercial products with a long history of use in construction

at ~700–750 °C. Although CRP results suggest that more widespread applications of geopolymers are possible, standards for the precursors and experience of process optimization are still needed along with long-term stability of these materials to be demonstrated. Research activity in non-nuclear applications of geopolymers is active and is likely to generate more knowledge of formulation and in particular, their long-term durability. Phosphate cements also receive attention for radioactive waste conditioning. Most of the experience in the nuclear field is with magnesium potassium phosphate cements. Although with limited use in the construction industry including rapid-set road and runway repairs and rock anchor bolt, these cements demonstrated persistence in the natural environment. Some novel cements have a history of commercial use and are known to be durable in a range of natural service environments. Standards are also available for C$A and CAC cements. Although used the geopolymer- and phosphate-type cements still require optimization with promising research results. Important to note that novel and conventional cementitious materials emplaced in the same repository have had no incompatibility problems identified.

The conclusion of CRP is that novel materials need a better benchmarking. Existing test methods may not give comparable results with different classes of materials.

4 Testing of Cementitious Wasteforms

The durability of cementitious materials is an important issue and their proper characterization using various methods plays a major role in the understanding of the use of such material for storage, transportation and disposal. The durability of

Table 1.3 Examples of techniques used within CRP for testing of cementitious materials

Property	Characterization tests
Composition	BET, DTA, XRD
Porosity	ASTM D 4284, gravimetrical methods, neutron imaging, mercury intrusion porosimetry (MIP)
Leaching	ISO method, ANSI/ANS-16.1-1986, Czech nuclear authority (SUJB) number 32811/2007
Sorptivity	Neutron imaging
Hydraulic conductivity/permeability	Permeability test (ASTM D 5084)
Pore size distribution	Neutron imaging, MIP
Thermal cycles	European standard EN 196-9
Micro structure and chemical composition	SEM, EDX, STEM, neutron imaging, acoustic emission, FTIR
Gas permeability	Torrent methods (ASTM C 577)
Specific gravity	ASTM C 642
Compressive strength	KS L 5105 (Korean standard), TEST N.278 (Slovakia)
Bulk density	KS F 2308 (Korean standard)
Biodegradability	Ukrainian requirements (NP 306. 608-95), GOST (Russia)

cementitious materials based on the designed specifications can only be quantified over long periods of research. However, due to time constrains, short-term characteristics such as the compression strength, tensile strength, porosity, sorptivity and permeability (leaching) are determined to be used in models aiming to determine the possible durability of these cementitious materials that satisfy waste acceptance standards. Examples of short-term characterization techniques used during the CRP are summarized in Table 1.3.

Most participants of CRP have used standard well-known methods for determining characteristics of cementitious materials in order to be used for durability calculations (see, e.g. [4]). Alternative characterization techniques for cementitious materials, such as neutron imaging and acoustic emission were demonstrated as possible techniques to complement the study of the durability of cementitious materials. The durability of cementitious materials is an important issue and the measuring thereof plays a major part in the understanding of the use of such material. Short-term characterization results can only be used as a guideline in the individual member states for wasteform development and durability predictions due to the fact that durability testing methodology is not standardized and scarce long-term durability research results on behaviour of cementitious materials under repository conditions.

Testing and the extrapolation of short-term test data to centuries (or longer) has proved to be very complex as concluded by CRP participants. Reliable data on the kinetics of all but the simplest processes are missing. It is suggested that quality assurance and criteria for acceptance have proven to be another problematical area. Although encouraging progress is reported on non-destructive testing (NDT), additional NDT test methods are required. Criteria for acceptance for final disposal need to be appropriate to the disposal site and activity of the package.

5 Modelling Long-Term Behaviour

Although it is generally accepted that cementitious materials can be engineered to perform over long times (based on analogy with natural, archaeological analogues and old structures), methods for quantifying such long-term service are not standardized and in some cases the need for durability evaluations is not recognized.

The durability of cementitious barriers can be assessed empirically by extrapolating short-term data. This approach offers limited opportunities for mechanistic analyses of processes that occur at different rates and sensitivity analyses of one-off conditions. Durability can also be assessed by reactive transport modelling for which parameters such as diffusivity and permeability (hydraulic conductivity) must be provided. This approach provides a mechanistic basis for durability and allows for analyses of interrelated and concurrent processes which can occur at different rates (fast processes addressed by the empirical approach and slow processes that may be relevant for the time frames required for LLW and especially ILW disposal facility performance). Reactive transport modelling can also support sensitivity analyses (one-off) and more rigorous uncertainty analyses applied to the overall disposal facility performance assessment or safety case evaluation. In addition, this type of modelling is suited for extrapolation and prediction of changes in physical and chemical properties of engineered cementitious materials (including waste forms) as a function of changing conditions over long times. Modelling durability has several aspects. Both conceptual models and mathematical expressions which represent the important processes that control performance (specifically flow and reactive transport models) are useful tools for analysing the level of understanding of the overall performance and for extrapolating performance beyond times for which observations and test data are available. Parameters used to describe and model leachability are needed for both, the cementitious matrix and also to radionuclides phases. Examples of the parameters used to describe the leachability in reactive transport models include: bulk chemistry, radionuclide concentrations, quantitative mineralogy, pore solution composition, leachate chemistry as a function of pH and Eh, and leachate-to-sample ratio. In addition, thermodynamic data and/or solubility data for the matrix phases are required. Thermodynamic data and sorption data for the phases and speciation of the radionuclides in water are also needed.

In a chemical approach to cementitious materials degradation under simple leaching conditions, four states of evolution of cements are suggested for benchmarking. These stages are achieved by sacrificial dissolution of cement substance as shown in Fig. 1.1 and are summarized as follows [3]: (1) Stage I: The pH is dominated by alkalis: all normal cement mineral hydrates are present. (2) Stage II: The pH is dominated by $Ca(OH)_2$: all normal hydrates are present. (3) Stage III: $Ca(OH)_2$ is consumed: C-S-H, depending on composition, pH is buffered in the range 10–12. (4) Stage IV: Only degradation and reaction products are left to condition pH. The four-stage approach is used in performance assessments for radioactive waste repositories for geochemical modelling of cementitious barrier

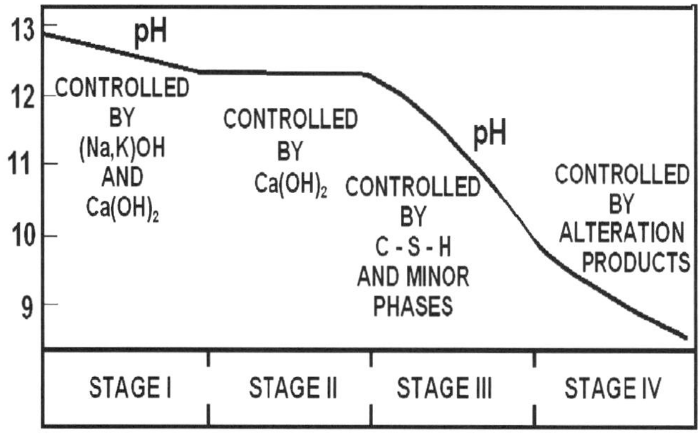

Fig. 1.1 Evolution of cements in a disposal environment

durability with respect to providing chemical buffering for radionuclide phases and also physical/hydraulic durability (porosity, permeability and tortuosity) of the matrix portion of the barrier. The objective is to estimate the life time of the buffer, for example how long the desirable high pH may persist, and to predict the evolution of the near-field chemistry, which is needed to estimate the corrosion rate of metallic containers and waste forms. The pH conditioning action depends partly on the persistence of a metastable phase, C-S-H, and more data are needed on its persistence as it may be subject to ageing as well as sacrificial dissolution. One-dimensional geochemical modelling can be used to evaluate the performance of coupled diffusive transport. This approach is often used to evaluate chemical performance of multilayer features encountered in disposal facilities. In the simple case, it is assumed that solute transport is by diffusion only, and assumptions are made for the geochemistry of the solid-solution system.

The Geochemist's Workbench, PHREEQC, GEMS and ORCHESTRA are examples of computer codes used with various thermodynamic databases to predict the long-term evolution of cementitious materials in the near-field environment. These codes are used to link the cementitious material degradation with radionuclide sorption and solubility parameters in the near-field subsurface environment.

Although numerous data are available on the release of radionuclides from various waste forms, modelling studies of the interactive processes of degradation in cementitious material are a much larger task than could be encompassed in this CRP. The local characteristics of the disposal site impose specific constrains on the evolution of barrier performance with time. Both simple and complex models are being developed: simple models as tools for semi-quantitative assessments and more complex and sophisticated models to couple processes, including reactive transport, with other chemical and physical degradation mechanisms.

The capacity to model all the effects involved in the dissolution and alteration of the waste form, in conditions similar to the disposal site, and to predict behaviours of cementitious materials used, is the final goal of the research undertaken by some research groups. As concluded during discussions modelling is still in its developmental stage. Nevertheless, it may be the way forward as compression of the time factor is difficult reliably to achieve by accelerated testing. However, it was also suggested that the database for modelling needs to be upgraded and enhanced and that modellers link to experimental studies designed to verify the key conclusions of mathematical models.

6 Conclusions

The CRP resulted in an active interchange of experiences among leading research groups on the behaviour of cementitious materials in long-term storage and disposal of radioactive waste. It has enabled to access valuable information on the underlying science and technology of cementitious materials used in radioactive waste management.

It was confirmed that cements are a suitable material for the immobilization of a variety of waste constituents due to their favourable chemical and physical properties. The cement hydration products formed favour sorption and substitution of cationic, anionic and neutral radwaste species into cement solids while the microstructure affords physical immobilization.

CRP concluded that interactions between a cement system and the waste stream have been shown to be complex. The research programmes of the CRP were carried out to fully understand and quantify these. Available research results have highlighted that, by controlling the internal chemistry, microstructure of the matrix and the hydration products formed, through incorporation of reactive admixtures, choice of curing temperature and water/cement ratio, systems may be developed selectively to enhance immobilization of a specific waste component or group of components.

CRP participants agreed that more research is required on the physical and chemical effects of waste ions on the cement structure during solidification: the formation of exotic hydration products, the speciation of waste ions and the resistance of the constituent solid phases to degradation. These interactions must be better understood because apparently slight changes in matrix chemistry could result in significant change in immobilization capacity and, if this capacity is reduced, accelerate release of radioactive material into the biosphere. Waste–cement interactions, and the time dependence of the immobilization potential, should be the focus of research.

Overall it is evidenced that an encouraging progress is reported in the CRP. While many tasks remain incomplete there is a greater understanding and definition of the remaining problems and of approaches and methodologies needed to solve them.

In addition to the report on achieved results of research, it is important to note that IAEA has facilitated creation of a user group or network of organizations that will continue to cooperate in this area of work. Bilateral agreements between participating organizations have been already established during CRP. It is expecting that cooperation among participants will continue outside of the CRP framework.

Acknowledgements Contributions of all CRP participants are acknowledged including leading investigators P.J. McGlinn, D. Perera, J. Govaerts, W. Bastiaens, R. Vicente, Li Junfeng, A. Vokal, A.M. El-Kamash, C. Cau-dit-Coumes, R.G. Yeotikar, D.S. Deshingkar, Joo-Wan Park, J.-H. Yoon, F. Dragolici, A.P. Varlakov, L.P. Sukhanov, I. Plećaš, D. Kicevic, M. Breza, P. Lichvar, A. Šajna, N. Zeleznic, W.C.M.H. Meyer, A. Waellisch, B. Torstenfelt, P. Mårtensson, L. Almkvist, F.P. Glasser, B.P. Zlobenko and C. Langton.

References

1. IAEA (2010) CRP on behaviors of cementitious materials in long term storage and disposal. http://www.iaea.org/OurWork/ST/NE/NEFW/wts_crp_cement
2. Drace Z, Ojovan MI (2009) The behaviours of cementitious materials in long term storage and disposal: an overview of results of the IAEA coordinated research programme. Mater Res Soc Symp Proc 1193:663–672
3. Glasser FP (2011) Application of inorganic cements to the conditioning and immobilisation of radioactive wastes. In: Ojovan MI (ed) Handbook of advanced radioactive waste conditioning technologies. Woodhead, Oxford, pp 67–135
4. IAEA (2007) Strategy and methodology for radioactive waste characterization. IAEA TECDOC-1537, IAEA, Vienna

Chapter 2
R&D on Cementation of Pulp with Complex Physical and Chemical Composition Stored in Tanks at the Mining Chemical Combine

L.P. Sukhanov, K.P. Zakharova, N.A. Naumenko, O.N. Pogorelko, A.V. Lebedeva, Yu. A. Revenko, V.A. Kravchenko, and V.I. Matselya

1 Introduction

Cementation is the basic technology for liquid LLRW/ILRW immobilization in the Russian Federation. In particular, the cementation technology will be implemented at the Mining Chemical Combine (MCC) in Siberia.

The object of cementation at MCC is "legacy waste," i.e., of radioactive hydroxide pulps (HP) have been accumulated in the MCC storage tanks. These pulps are a result of previous activities of the MCC radiochemical plant. The volume of pulp—several thousands of cubic meters.

The MCC radiochemical plant is to be decommissioned in the near future. Accumulated pulps will be removed from the tanks and sent for cementation.

The aim of this report is to provide brief information about R&D results relevant to the above issue.

The management of pulp includes the following stages:

Stages being currently implemented

- Removal of liquid phase from storage tanks.
- Dissolution of pulp solid phase.
- Extraction of U and Pu from solutions (if appropriate).
- Collection of undissolved pulp residues in interim tank (as "secondary" pulp).

Stages planned

- Immobilization of undissolved pulp residues ("secondary" pulp) in a cement matrix.
- Long-term storage/disposal of cement compound.

L.P. Sukhanov (✉) • K.P. Zakharova • N.A. Naumenko • O.N. Pogorelko • A.V. Lebedeva
A.A. Bochvar High-Technology Research Institute of Inorganic Materials, Moscow, Russia
e-mail: lpsoukh@bochvar.ru

Yu. A. Revenko • V.A. Kravchenko • V.I. Matselya
The Mining Chemical Combine, Krasnoyarsk Region, Russia

F. Bart et al. (eds.), *Cement-Based Materials for Nuclear Waste Storage*,
DOI 10.1007/978-1-4614-3445-0_2,

The use of cementation technology for MCC pulp immobilization is based on different factors: ability to hold U and Pu in a cement compound, possibility to enhance ^{137}Cs confinement in a cement compound using sorbent additives, the process is carried out at low temperature and does not require sophisticated equipment, availability and relatively low cost of matrix materials, and absence of secondary waste.

^{137}Cs and ^{239}Pu in pulp are "critical" radionuclides for safety of long-term cementitious pulp storage. Safety of cement compound long-term storage/disposal facility is based on several barriers: matrix composition (cement + sorption additives), container (200-L drums) and packaging elements, engineering barriers of the storage/disposal facility (building structures, buffer materials, and sealing elements), massive rock materials between the storage/disposal facility and environment (the facility is located at a depth of ~200 m below the surface).

Main components of pulp sent for cementation include Fe_2O_3, Fe_3O_4, AlO(OH), Cr_2O_3, MnO_2, SiO_2, Ni and Sr and Cs (as hydro aluminum silicates), PuO_2 as silicates, aluminates, aluminum silicates, U as polyuranates.

Content of solid phase in the pulp is 100 ± 60 g/L, β-activity (mainly ^{90}Sr) is 1 ± 0.5 Ci/kg.

Pulps have multicomponent chemical composition with significant content of long lived alpha-radionuclides. Some components of the pulp solid phase can react with cement components (hydrated oxides of Fe, Al, Mn, Ni, and Cr).

Main objectives of investigations were to obtain experimental data to make recommendations on production of cement compounds satisfying the national waste acceptance requirements and to check possibility for using the facility with a pulse mixer for pulp cementation (new type of a cementation facility).

Scope of research investigations includes determining the characteristics of the produced cement compound (mechanical strength, water resistance, frost resistance, flowability, heat release, loading capacity of waste and others), assessing behavior of cement compounds during their production, and short-term storage and testing of the facility equipped with a pulse mixer for cementation of MCC pulp.

Main investigations have been performed using the simulated radionuclide-containing pulp. A set of experiments performed using the real pulp samples has been conducted at the Central radiochemical plant laboratory of MCC. All facilities and equipment available included pilot cementation facilities, laboratory equipment, radiation control instruments, and other equipment used at A.A. Bochvar Institute and MCC.

2 Procedure of Experiments

In laboratory conditions at VNIINM, there has been a wide range of experiments carried out using the hydroxide-pulp simulators. The experiments were meant to develop optimal parameters for the cementation process to ensure obtaining the required quality cement compound.

Table 2.1 Composition of hydroxide pulp simulators

	Components of undissolved pulp residue (wt%)	
Pulp component	Pulp simulator No. 1	Pulp simulator No. 2
$Al(OH)_3$	11.3	9.1
$Fe(OH)_3$	27.8	22.5
$Cr(OH)_3$	2.3	1.9
$Mn(OH)_2$	38.1	30.8
$Ni(OH)_2$	14.2	11.4
SiO_2	6.3	24.3

In the course of the research investigations, the following characteristics indices were determined: mechanical strength (after 7, 14, 28, 56, and 90 testing days), water resistance and frost resistance of a cement compound, radionuclide leaching rate, flowability, and specific heat release during hydration of compounds.

Two types of pulp simulators were used containing basic solid-phase macrocomponents and differing in the hydroxide-to-silicon oxide ratio (Table 2.1).

The following binding materials and sorption additives were used to prepare cement compounds:

- Portland cement PC 500-D0 (CEM I 42.5B).
- Slag Portland cement SPC 400 (CEM III 32.5).
- Bentonite clay M4T1K.
- Clinoptilolite (Kholinsky deposit, Siberia).
- Superplasticizer C-3.

A cemented pulp composition, a binder type, a cement-compound loading capacity, and a water-to-binder ratio (W/B) were varied in the course of the research. The cement compound samples, 2 × 2 × 2 cm in size, were placed in a humid-air atmosphere for hardening.

A set of experiments using the real pulp samples has been conducted at the Central radiochemical-plant laboratory of MCC to check VNIINM's recommendations. To carry out these investigations, the pulp sample with a volume of about 2 l was taken from a storage tank. The sample represented an inhomogeneous mass of sintered particles stuck together. The density and humidity of the pulp were, respectively, 1.55 g/cm^3 and 51.3 %. With the W/B ratio ranging from 0.7 to 0.9, the degree of the compound filling with the pulp was 4.5 %, 9 %, and 40 %.

3 Experiment Results and Their Discussion

3.1 Laboratory Investigations

The results of some laboratory experiments are shown in Figs. 2.1, 2.2, 2.3, 2.4, and 2.5.

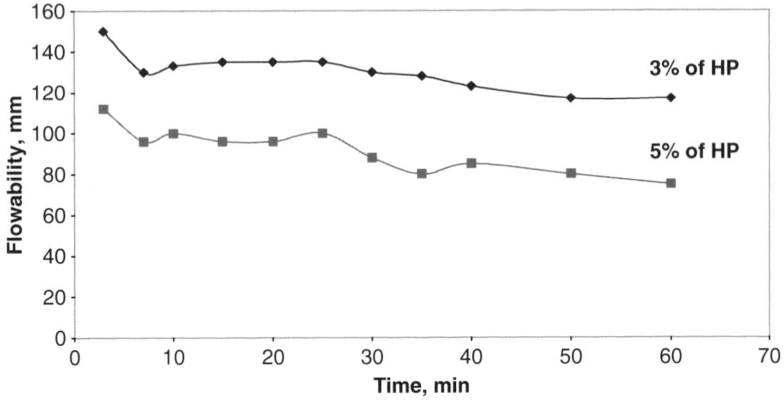

Fig. 2.1 Effect of loading capacity on cement paste flowability (simulated pulp)

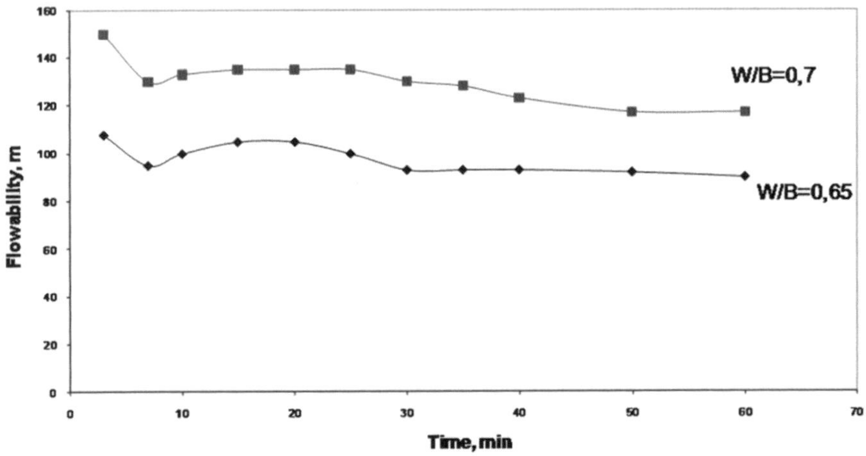

Fig. 2.2 Effect of water/binder ratio on cement paste flowability (simulated pulp)

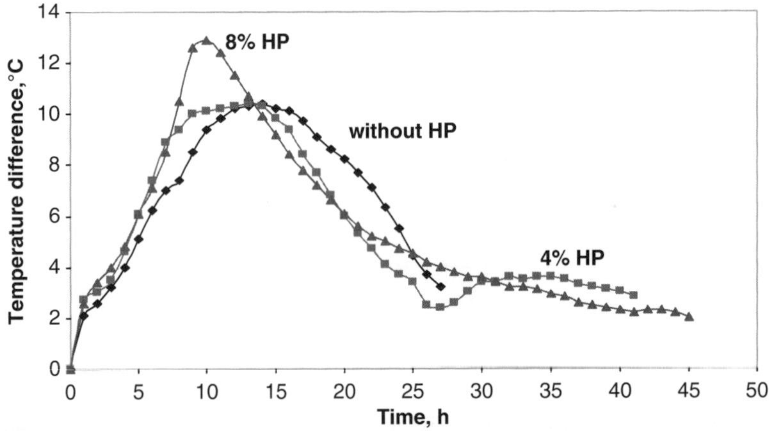

Fig. 2.3 Effect of pulp content in cement compound on heat generation (simulated pulp)

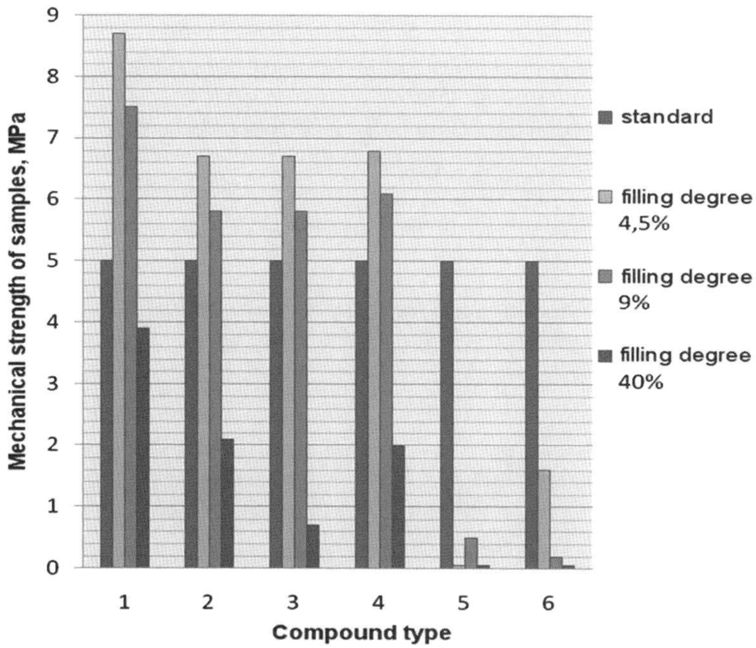

Fig. 2.4 Effect of loading capacity (filling degree) on mechanical strength of cement compound (real pulp from storage tank)

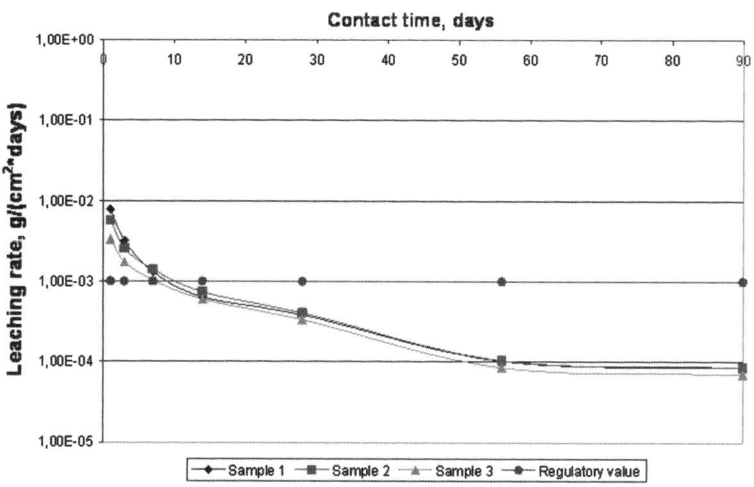

Fig. 2.5 Cesium-137 leaching tests for period 90 days (real pulp from storage tank)

Laboratory research results

- Portland cement and slag portland cement with sorbent additives (klinoptilolite or bentonite) may be used as a matrix material for MCC pulp immobilization.
- The cement compound had the required quality on mechanical stability, water resistance, freezing resistance, and leachability.
- Experiments helped to determine optimal parameters for cementation process and to make preliminary recommendations on technological regime of pulp cementation.

3.2 Testing of Pulse Mixer for Pulp Cementation

A.A. Bochvar Institute has developed the innovative cementation facility with a pulse mixer. The facility has the following advantages: there is no stirrer in it and cement compound is guaranteed to be discharged from the mixer. The testing objectives were to check (1) possibility to get homogeneous cement compound of a required quality and (2) preliminary technological recommendations made according to the laboratory research results.

Testing of pulse mixer description includes the following (Fig. 2.6). The cementation facility mixer was loaded with pulp simulators and a mixture of PC (or SPC) and bentonite (10 wt%), which was added to the mixer by permanent stirring for 20 min. Once completely loaded, the mixer was stirred for 5 min and then discharged. Samples were drawn in the beginning, in the middle and in the end of the discharge process to determine the flowability of the compound. The mechanical strength (after 7, 14, 28, 56, and 90 days), water resistance, freezing resistance, and leaching rate of the obtained cement compound were determined.

Pulse mixer testing showed

- Using the pulse mixer for pulp cementation makes it possible to produce a homogeneous cement compound with flowability allowing for the compound discharge from the mixer to the drum.
- The obtained cement compound had the required quality.

Based on results of the experiments, the recommendations have been provided on the cementation process parameters and characteristics of the cement compound obtained.

4 Conclusion

The obtained experimental data allowed to make preliminary technological recommendations on production of cement compounds that satisfy the national waste acceptance requirements.

The next step will be creation of full-scale pilot cementation facility and its testing using simulated pulp and real MCC pulp.

Fig. 2.6 Cementation pilot facility with pulse mixer

Chapter 3
Cementation in High Energy Mixer of ILW Surrogate Slurry. Demonstration of the Process at Semi-industrial Scale

G. Cardinal, P. Regnard, E. Tronche, and L. Donnet

1 Description of the Platform

The platform is similar to the future cementation workshop of the Liquid Effluent Treatment Station (STEL) at Marcoule. The purpose of the Liquid Effluent Treatment Station is to receive active liquid effluent, decontaminate it by means of a coprecipitation treatment, and then contain the radioactive slurry obtained in a matrix that complies with the outlet constraints selected. By 2015, the renovation of the Liquid Effluent Treatment Station will make it possible to replace the current bituminization conditioning procedure with a cementation procedure (Fig. 3.1).

A pilot cementation platform was installed in 2009 in order to validate the procedure at an intermediate scale between that of the laboratory and the future nuclear facility. The purpose of the test platform is to validate the choice of mixing technology, to adapt this commercial system for the nuclear industry, and to verify the compliance of the waste container produced. The platform's layout over three levels is similar to that of the future nuclear facility (Fig. 3.2).

1.1 Upper Level of the Platform

The upper level is specifically designed for the storage and batching of different components of the cement mix.

The dry materials such as cement and sand are stored in weighing hoppers. There is one batcher for cement and a second batcher for sand. These materials are introduced into the mixer by the force of gravity.

G. Cardinal (✉) • P. Regnard • E. Tronche • L. Donnet
Commissariat à l'Energie Atomique et aux Energies Alternatives, DEN/MAR/DTEC/SDTC, Bagnols-sur-Cèze, France
e-mail: gilles.cardinal@cea.fr

F. Bart et al. (eds.), *Cement-Based Materials for Nuclear Waste Storage*,
DOI 10.1007/978-1-4614-3445-0_3, © Springer Science+Business Media New York 2013

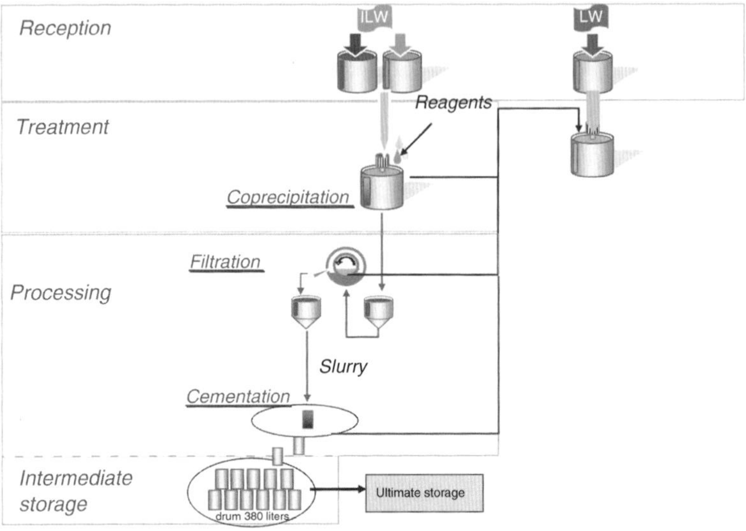

Fig. 3.1 Operation of the Liquid Effluent Treatment Station (STEL) at Marcoule

Fig. 3.2 Cementation platform

The adjuvants, superplasticizer for example, are introduced by a volumetric proportioning pump.

The waste, a non-radioactive slurry in this case, is introduced into the mixer by means of a peristaltic pump. Batching takes place by weighing the drum containing the slurry.

1.2 Intermediate Level: The Mixer

1.2.1 Choice of Technology

Feedback on mixer technologies was produced by French Atomic Energy and Alternative Energies Commission (CEA) facilities and other entities from the nuclear energy industry. A number of important points for the choice of technology to be implemented were revealed:

- The power dissipated by the stirrer and the hydrodynamic regime obtained depend on the position and geometry of the stirrer. These parameters are of primordial importance, especially in the case of very viscous cement mixes.
- A gravity drainage system facilitating the flow of the mixture is favored.
- Driving the stirrer from the bottom of the vessel should be avoided in order to limit maintenance of the leaktight seal system and to facilitate the cleaning of this area which has retention points.
- The vessel and stirrer rinsing system must be efficient and consume little water.

Taking these criteria into account led CEA to choose an Eirich brand mixer (Fig. 3.2). This type of machine has the advantage of having both a stirrer and a rotating vessel. This is advantageous for highly viscous mixes that need high mixing power. The mechanical maintenance of the stirrer is performed from above, removing the leaktightness problem.

Drainage takes place by the force of gravity through a drainage gate in the center of the base of the vessel. The stirrer and fixed scraper direct the material towards the center facilitating the drainage of the mixer. In the event the hydraulic drainage gate opening system were to dysfunction, opening can be forced using a manually operated pump. Similarly, a loss of pressure in the hydraulic circuit causes the drainage gate to open due to the effect of the weight of cement mix present in the mixer.

The high-pressure rinsing system comprises a retractable pipe fitted at the end with a rotating nozzle for the mixer vessel and fixed nozzles for the drainage hopper. Water consumption for cleaning the machine is to be consolidated. Specifically, the notion of the cleanness of the mixer between an application for the nuclear industry and a standard industrial application is different.

The mixer and the instrumentation and control station are installed at the intermediate level. The main characteristics of the R09 mixer installed on the technological platform are as follows:

- The vessel has a volume of approximately 350 L (900 mm; H 580 mm) and a useful volume of 150 L (or 240 kg). According to Eirich, this size of mixer is representative up to a useful volume of 1,000 L.
- The other functions described in the introduction are present on this model.
- This mixer is half the scale of the future Liquid Effluent Treatment Station mixer.

1.3 Lower Level: Pouring Station

The pouring station is at the lower level. The stainless steel container docks under the mixer drainage hopper. These dimensions are, respectively, a diameter of 685 mm and a height of 1,000 mm. The volume is 380 L.

1.4 Laboratory

Samples of the mix taken from pouring are characterized in the laboratory attached to the platform. Characterization of the fresh mix involves the measurement of specific density, viscosity, then the maximum temperature during setting and the associated time, as well as the hydration heat measured by semiadiabatic Langavant calorimetry.

2 Description of a Cementation Test

The cementation formulation was adjusted as follows:

- Sand with a D50 of 593 μm was selected in order to obtain a pourable cement mixture. It is non-alkali-silica reactive. The content in the mixture is of the order of 24 % pds.
- A CEM 1 cement with a low hydration heat was selected. The cement used was a Portland cement that is resistant to sulfate attack with a batching of 670 kg/m^3, which is approximately 37 % weight.
- The plasticizer is BASF Glenium 51 following the Plasticizer/Cement batching comprised between 1 and 24 % weight.
- The non-radioactive slurry used for the tests has similar chemical and rheological characteristics to those of real waste (Fig. 3.3). The initial effluent was reconstituted on the basis of the average chemical composition of the actual

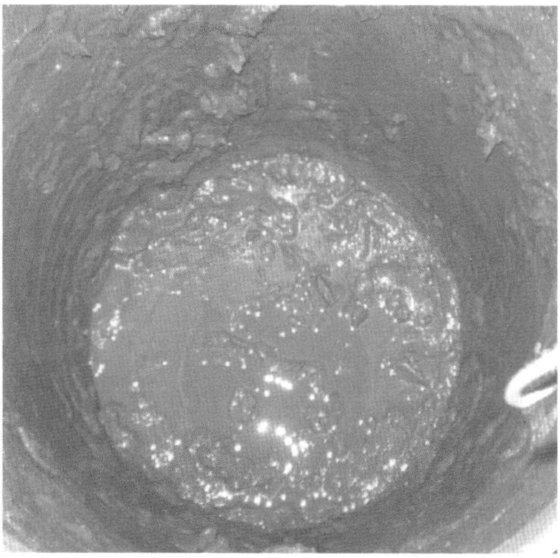

Fig. 3.3 Reconstituted slurry

effluent received in the Liquid Effluent Treatment Station (STEL). Next, the manufacturing process follows the same treatment protocol as that used in the STEL at Marcoule. During the first stage, the coprecipitation reagents are introduced into the vessel of effluent. Then the suspension obtained is filtered through a rotary filter with a precoating layer of perlite. The dry slurry extract obtained is of the order of 200 g/kg. The rate of incorporation equivalent in dry slurry in the cement matrix is 8 % weight.

The test procedure in terms of the order in which the constituents were introduced, and the phasing of the various stages of mixture, is described in Table 3.1. The mixing parameters were determined initially with the supplier and then optimized on the basis of the results of the tests performed on the platform. The slurry is introduced after the dry material in order to obtain a good mix of the dry loads beforehand and also for safety reasons. In the event that the sand or cement batching system dysfunctions the slurry introduced in the mixer would be drained into a container. The treatment of this non-compliant container would be difficult to manage.

Four successive concrete mixes were necessary to fill a 380 L drum. The total duration of the production of the four concrete mixes by the platform was approximately 2 h 30 min (Fig. 3.4).

The retention of material in the mixer increases respectively by 30 kg in the first concrete mix to 75 kg after the last concrete mix. The cycles of power consumed by the stirring tool (Fig. 3.4: medium grey curve with squares) are similar for the various concrete mixes. When the dry materials are introduced, the power consumed increases up to a steady state of approximately 4.5 kW. There is a power surge (approximately 6 kW) when the slurry is introduced that then falls rapidly and stabilizes again around 4.5 kW. The power surge is smaller from one concrete mix to the next as the residual mass of the hold-up is already homogenized. The temperature inside increases by 6 °C during the first concrete mix and then it

Table 3.1 Protocol for production of a cement mix container by the pilot mixer

Stage No.	Phases	Speed of the mixing tool expressed as a percentage of V_{max}	Speed of the vessel expressed as a percentage of V_{max}	Duration expressed in *s*
1	Incorporation of the sand	44	50	60
2	Incorporation of the cement	44	50	300
3	Mixing	44	50	30
4	Mixing	66	50	30
5	Mixing	44	50	10
6	Incorporation of the slurry	44	50	360
7	Mixing	44	50	30
8	Mixing	66	50	60
9	Incorporation of the superplasticizer	30	30	20
10	Mixing	44	50	180
11	Drainage preparation	50	75	10
12	Draining	100 (≈400 tr/min)	100 (≈40 tr/min)	20

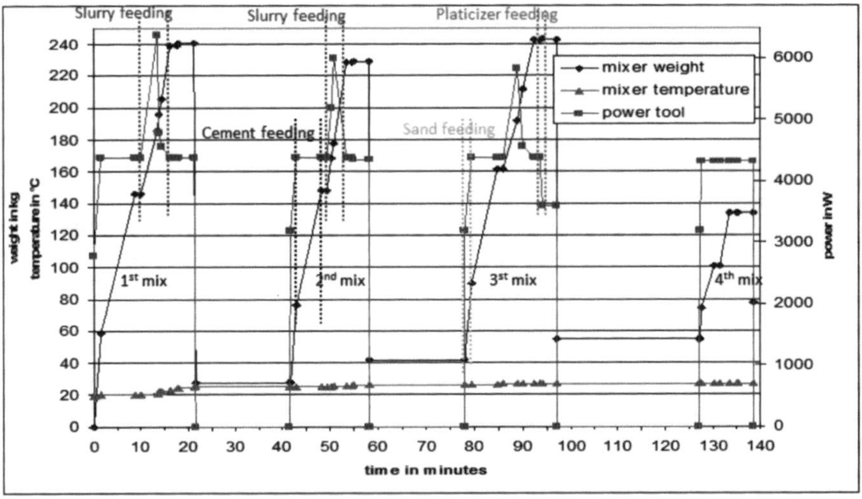

Fig. 3.4 Diagram of container production

remains constant for the rest of the test. The introduction of materials at ambient temperature (20 °C) probably stabilizes this slight increase in temperature.

All of the internal surfaces, i.e., the mixer vessel and the drawdown hopper, are covered with a film of fairly thick mix.

The repair of the hold-up is as follows:

- The area cleaned by the scraper, i.e., the side wall and bottom, is covered with a regular film approximately 1.5 mm thick.
- The mixing tool is also covered with a regular film approximately 1.5 mm thick.
- The scraper is covered with a film that is 2–3 mm thick. In addition, mix agglomerates are present behind the scraper on the horizontal part and on the vertical part.
- The cover of the mixer is covered with a 2–3-mm-thick film. There is mix spattering inside the mixer feed points to a height of approximately 30 cm.
- The drainage hopper and the drainage gate are covered by a film that is 2–3 mm thick. Agglomerates were observed in the gate opening/closing mechanism and in the corners of the hopper.

Technically, this test demonstrates the feasibility of successively performing several concrete mixes without cleaning the mixer. The leaktightness of the drainage gate was preserved.

The mixer supplier indicated a volume of 30–50 l for cleaning the mixer. In addition, operators may need to clean the mixer manually from time to time.

The tests show that complete cleaning of the mixer needs a volume of water of approximately 130 l. They also show that a series of short rinse cycles is more effective than a single rinse consuming an identical volume of water. However, these tests revealed shortcomings in the automatic cleaning system:

- The mixing tool creates a shadow zone on the cover of the mixer that is not cleaned. The geometry of the top of the scraper is difficult to clean.
- Cement mix spattering was present in the mixer feed points.
- The rectangular section of the drainage hopper makes the cleaning of 90° angles difficult. The positioning of the nozzles needs to be reviewed.

3 Characterization of the Cement Matrix

The Water/Cement ratio of the concrete mixes varies between 0.85 and 0.95. This fluctuation is related to the variation of the dry extract of the slurry between 199 g/kg and 206 g/kg.

The dynamic viscosity of the cement mixture measured after each of the four concrete mixes increased from 1,600 mPa s to 2,500 mPa s. On the other hand, the W/C ratio increases from 0.85 to 0.95. An increase in the water content should, in theory, lead to a more fluid cement mix. One assumption—to be confirmed—may be put forward. The rate of recovery of material after the first drainage is 88 %. The first germinations of hydrates in this material occur while the next are being created. The formation of these hydrates could be a cause of the increase in the viscosity of the mixture. As the viscosity increases, the retention of the material in

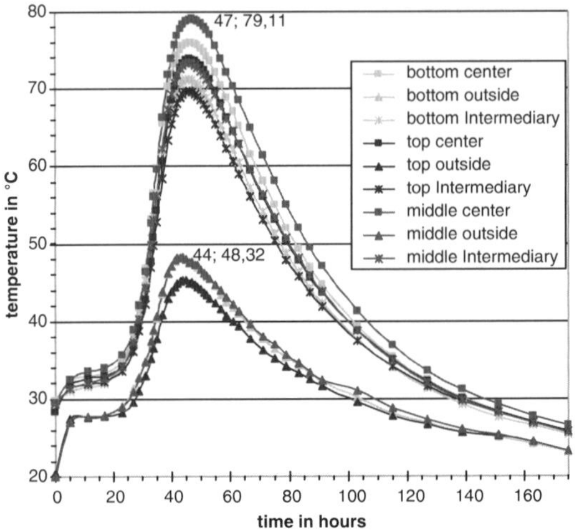

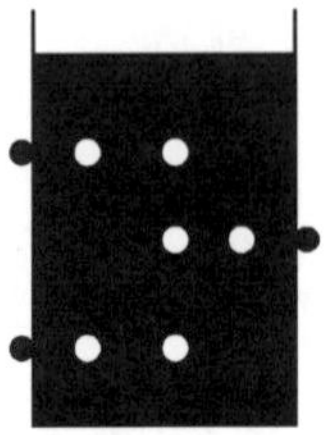

Placement of the thermocouples in the container.

Fig. 3.5 Diagram showing the evolution of the temperatures in the container during setting

the mixer increases, respectively, from 30 kg at the first concrete mix up to 60 kg after the last concrete mix.

The specific density value measured in situ on the fresh mix is 1.82. The inferred value of the measurement of porosity accessible to water in test specimens at 90 days is similar and is 1.85.

The temperature when the mix sets is monitored by Langavant semi-adiabatic calorimetry and in the waste container.

The container is fitted with nine thermocouples distributed as shown in Fig. 3.5.

The maximum temperature at the core of the container is 79.1 °C. The temperature of the wall is 48 °C. The maximum temperature measured in the calorimeter is 59.4 °C (Fig. 3.6). The increase in temperature is 20 °C compared with the temperature measured using the calorimeter.

The setting time is within a period of 45–50 h. No exudation of water on the surface of the container is observed after 24 h.

The porosity accessible to the water is homogeneous over the entire container and of the order of 50 %. The resistance to compression on the test specimens 11 × 22 cm at 90 days is between 20 and 25 MPa. They are clearly greater than the 8 MPa required by ANDRA. The appearance of the mix is homogeneous. Some flaws (porosities 1 cm in size) were observed on the edges of the container. At microscopic level, small clusters of slurry are distributed homogeneously throughout the container. The cement matrix incorporates these slurry clusters (Fig. 3.7).

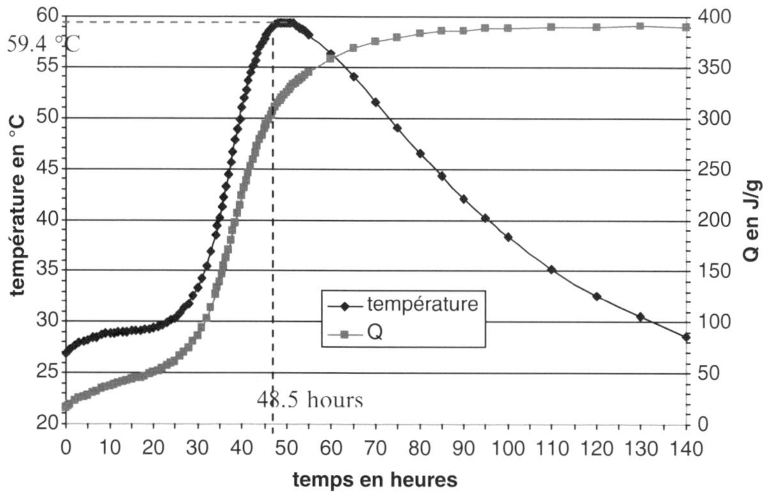

Fig. 3.6 Semi-adiabatic calorimetry of the mix

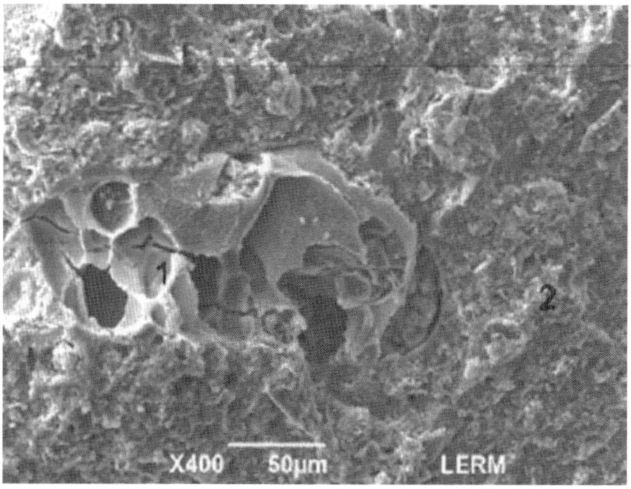

Fig. 3.7 Scanning electron microscopy—appearance of the freshly fractured matrix (1: slurry cluster; 2: hydrated matrix)

The results of the characterizations performed in the laboratory comply with those obtained during determination of the cement formulation at laboratory scale. The production of this drum makes it possible to comprehend the increase in temperature due to the volume effect. This increase to the hottest point is of 20 °C in relation to the laboratory tests.

4 Conclusion

The half-scale tests on this platform make it possible to validate the results obtained in the laboratory on the cement matrix. The influence of the volume of the container on the maximum setting temperature was revealed and is a 20 °C increase in temperature in relation to the laboratory tests.

R&D will make it possible to optimize the mixing parameters, validate the modifications of the machine to limit retention of the cement mix, define a nominal operating range, and also consider transient and impaired conditions.

Part II
Physico-Chemical Processes Occurring in Cement-Waste Forms at Early Age

Chapter 4
Hydration of Blended Cements

B. Lothenbach

1 Introduction

The use of supplementary cementitious materials (SCMs) such as blast furnace slag or fly ash represents a viable alternative to Portland cements and utilizes by-products of industrial manufacturing processes. An industrial application of SCMs, however, is often hindered by the different and varying chemistry and that other hydrates are formed during hydration than in Portland cements. In Si-rich systems, portlandite can be absent and a calcium silicate hydrate phase (C-S-H) with a low Ca/Si ratio develops. A lower Ca/Si ratio leads also to an increase in alkali uptake by C-S-H [2] and to a reduction of pH values in the pore solution [5]. Most of the available studies on the properties of blended systems focus on mechanical or on durability aspects of a specific silica fume, fly ash, or slag, while little is known about the fundamental connection between the composition and hydrates formed as well as its impact on the long-term development of such systems. In this chapter, the hydration of a Portland cement and of two blended cements is investigated experimentally. Thermodynamic modeling is used to predict the changes during hydration and the changes associated with the presence of increasing amounts of SiO_2.

B. Lothenbach (✉)
Empa, Lab. Concrete & Construction Chemistry, 8600 Dübendorf, Switzerland
e-mail: barbara.lothenbach@empa.ch

F. Bart et al. (eds.), *Cement-Based Materials for Nuclear Waste Storage*,
DOI 10.1007/978-1-4614-3445-0_4, © Springer Science+Business Media New York 2013

2 Materials and Methods

2.1 Materials

The experiments were carried out at 20 °C using three different mixtures: a Portland cement CEM I 42.5 R HS (*OPC*, $w/b = 0.8$), a shotcrete mixture (*ESDRED*, $w/b = 0.5$), and a low-alkali cement (*LAC*, $w/b = 1.1$). The shotcrete ESDRED consisted of a 60 % CEM I 42.5 N, 40 % silica fume, an alkali-free accelerator, and a polycarboxylate-based superplasticizer. To prepare the low-alkali cement LAC, 90 wt% CEM III/B 42.5 L was mixed with 10 wt% nanosilica and a 2.6 g/100 g of superplasticizer, resulting in 30 % PC, 60 % blastfurnace slag, and 10 % SiO_2.

The composition of the three cements and the slag was determined by XRF. The mineralogical composition of the unhydrated cement was quantified by Rietveld analysis according to the methodology presented in Le Saout et al. [6] and presented in Table 4.1. The amorphous amount of the anhydrous sample was calculated using an internal standard.

Table 4.1 Mineralogical composition of the cements

	Weight fractions (wt%)		
Phases	CEM I 42.5 R HS (*OPC*)	CEM I 42.5N (*ESDRED*)	CEM III/B 42.5L (*LAC*)
Alite	52.7	52.8	16.7
Belite	14.4	22.8	5.6
Aluminate	1.1	5.0	0.9
Ferrite	17.2	7.9	3.4
Periclase	3.3	1.0	0.3
Calcite	1.7	2.9	3.9
Quartz	–	0.5	0.3
Anhydrite	1.0	1.9	–
Hemihydrate	2.2	1.7	–
Gypsum	3.8	–	2.2
Syngenite	–	2.0	–
Dolomite	2.5	1.5	0.7
Amorphous	–	–	65.9
Measured composition of the slag (wt%)			
CaO	–	–	42.9
SiO_2	–	–	34.9
Al_2O_3	–	–	11.1
Fe_2O_3	–	–	0.3
MgO	–	–	6.7
Na_2O	–	–	0.3
K_2O	–	–	0.4
S(-II)	–	–	1.0

2.2 Hydration Experiments

The composition of the hydrating OPC, ESDRED, and LAC cements was investigated in paste samples from 1 h to 1,310 days. For each of time step, an individual cement paste sample was prepared consisting of 1 kg cement. The appropriate amount of water and admixtures was mixed twice for 90 s according to EN 196-3. The pastes were cast in 0.5 and 0.1 L PE-bottles, sealed (to exclude the ingress of CO_2) and stored at 20 °C until they were sampled. After different hydration times, solid samples were cut from the 0.1 L bottles and examined immediately by XRD. X-Ray Diffraction (XRD) data are collected using a PANalytical X'Pert Pro MPD diffractometer. For SEM/EDX analysis, the samples were dried at 40 °C, pressure impregnated with epoxy resin, polished and carbon coated. BSE and image analysis (SEM-IA) was used to determine the amount of unreacted slag based on the gray level histogram and morphological segmentation. The reaction of the anhydrous slag was determined by comparing the volume of anhydrous slag in the hydrated pastes with the volume of the slag content in the unhydrated paste.

Pore fluids of the hardened samples were extracted from the 0.5 L samples using the steel die method and pressures up to 530 N/mm^2. The solutions were immediately filtered using 0.45 μm nylon filters. The pH measurements were carried out immediately after filtration using small aliquots of the pore solution; the concentrations of other ions were analyzed using ICP-OES.

2.3 Thermodynamic Modeling

Thermodynamic modeling was carried out using the geochemical modeling code GEMS [4], which computes equilibrium speciation of the dissolved species as well as the kind and amount of solids precipitated. The thermodynamic data for aqueous species as well as for many solids are taken from the PSI-GEMS thermodynamic database [3], while data for cement minerals are taken from the recent cemdata07 database [7]. The cemdata2007 dataset includes thermodynamic data of common cement minerals such as C-S-H, different AFt and AFm phases, hydrotalcite, and hydrogarnets. No restrictions on the kind of hydrates calculated were imposed, with the exception of siliceous hydrogarnet ($C_3AS_{0.8}H_{4.4}$), whose formation was suppressed as its formation seems to be kinetically hindered at ambient temperatures. For modeling, the formation of ideal solid solution between Al- and Fe-containing analogues is assumed, such that the expressions ettringite, monocarbonate, strätlingite, or hydrotalcite in this chapter all refer to the solid solution between the Al- and Fe-containing analogues. As slag is present in the LAC mixture, the formation of FeS (troilite) is taken into account.

3 Hydration

3.1 Experimental Results

The blending of Portland cement with mineral additions with high silica content such as blastfurnace slag, fly ash, and/or silica fume influences the hydrate assemblage of the hydrated cement. The processes during the hydration of OPC are well known. The main hydrate phases of the OPC studied included C-S-H (calcium silicate hydrate), portlandite, ettringite, hydrotalcite, and monocarbonate. The pH value of the pore solution was 13.4 as shown in Fig. 4.1. The hydrate assemblage and the composition of the pore solution remained more or less stable after longer hydration times.

During the first day, the hydration of the ESDRED cement (60 % PC, 40 % silica fume, alkali-free accelerator) proceeded similarly to the OPC hydration. The gypsum present in the cement was dissolved after 1 day and most of the clinkers were hydrated after 1 week (Fig. 4.2). A significant quantity of ettringite was formed during the first minutes due to the presence of aluminum and sulfate in the alkali-free accelerator. The presence of small amounts of portlandite was observed between 1 and 7 days of hydration and the presence of hemi- and/or monocarbonate up to 56 days. After 1 year, the main hydration products that could be identified by XRD and TGA were ettringite and tobermorite-like C-S-H. TGA indicated the decrease of the amount of pore solution during hydration and the formation of approx. 25 g ettringite/g hydrated ESDRED. The 40 wt% silica fume originally present was found to react only slowly so that after 3.5 years of hydration, 10 wt% silica fume had not yet reacted. The SEM/EDS analysis indicated a mean Ca/Si ratio of 1.3 ± 0.2 and an

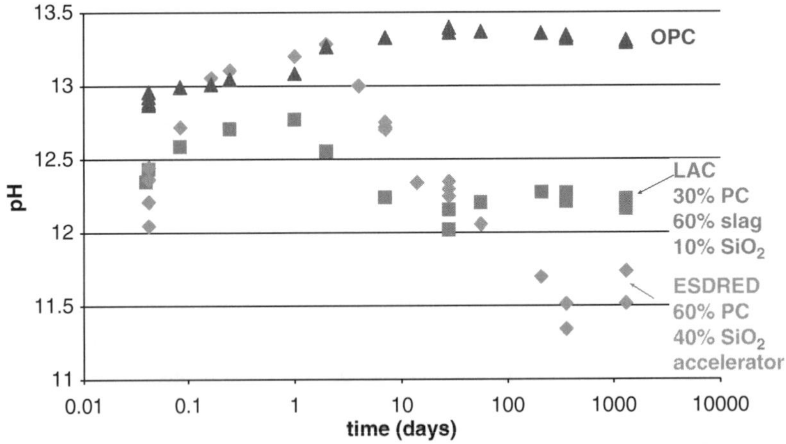

Fig. 4.1 Measured pH values in the pore solution of OPC, LAC, and ESDRED pastes

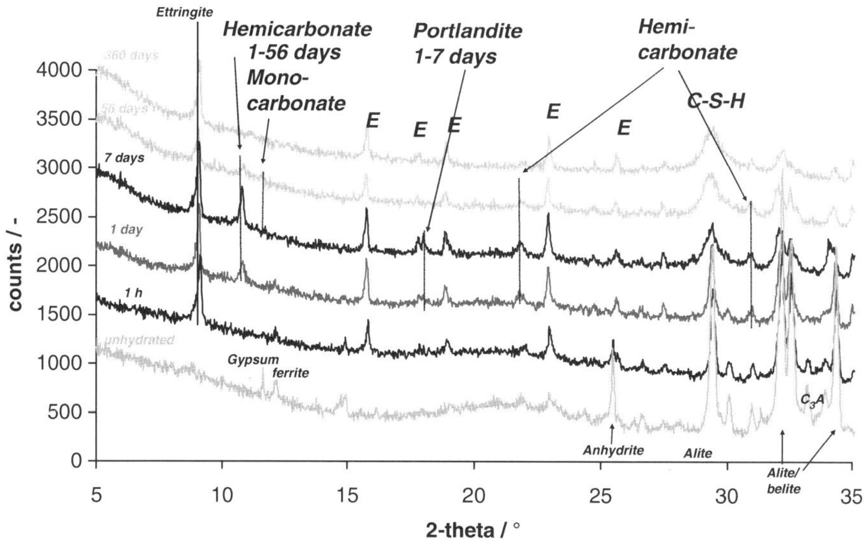

Fig. 4.2 XRD of the hydration of ESDRED (60 % PC, 40 % silica fume, alkali-free accelerator). *E* ettringite

Al/Si ratio in the C-S-H of 0.05–0.1. The pore solution had a pH of 11.5 (Fig. 4.1), somewhat lower alkali concentrations than in an OPC system, but a relatively high Ca concentration of 30 mM.

The hydration of LAC (CEM III/B plus 10 % nanosilica) proceeded similarly. C-S-H (Ca/Si = 1) and ettringite were the main hydration products, while no portlandite was observed. EDX indicated the presence of hydrotalcite, and saturation indices calculated from the pore solution show the presence of strätlingite after a few months (Lothenbach et al. submitted). While the nanosilica had reacted within the first days, the slag dissolved relatively slowly. However, after 1 year 90 % of the slag had reacted and the SEM/BSE pictures of the hydrated samples showed only the presence of a few unhydrated slag particles. A relatively dense microstructure was visible with only little coarse porosity. The pore solution after a hydration time of 1 year was dominated by sodium, potassium, hydroxide, as well as by the different sulfur species. The high fraction of sulfide indicates reducing conditions. A pH of 12.3 was determined after one year and longer (Fig. 4.1).

3.2 Modeling

Quantification of the phase assemblage is difficult in silica-rich blends as an important part of the starting material and the C-S-H, the main hydration product, are X-ray amorphous. Thus, thermodynamic modeling offers an alternative to

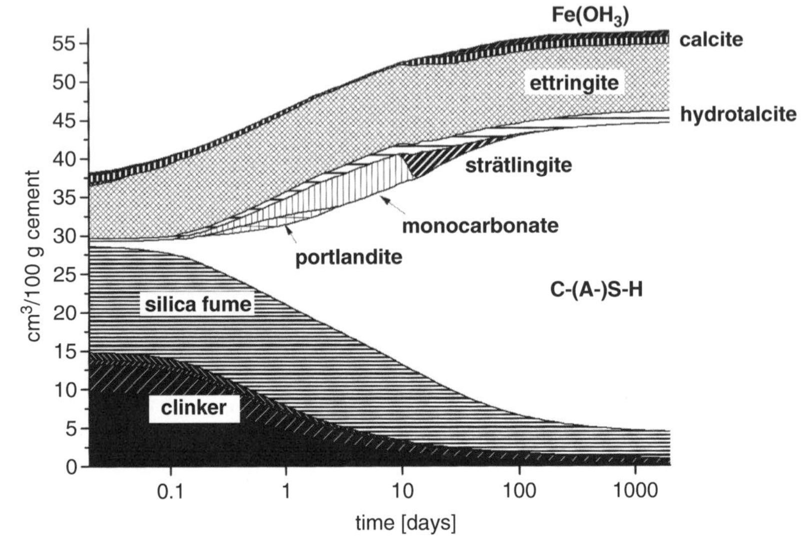

Fig. 4.3 Modeled changes during the hydration of ESDRED expressed as volume (cm^3/100 g unhydrated cement)

investigate changes during the hydration and to predict quantitatively the hydrate assemblage as illustrated for the silica fume blended ESDRED cement in Fig. 4.3. The thermodynamic calculations are based on an empirical model to describe the progress of hydration [8, 11], on the measured progress of the slag and silica fume reaction, and on the composition of the starting materials. Based on the EDX measurements an Al/Si ratio in the C-(A-)S-H of 0.1 was assumed.

The thermodynamic modeling of the shotcrete ESDRED predicts the formation of C-S-H with Ca/Si ratio of ~1, ettringite, and some small quantities of hydrotalcite (Fig. 4.3). During the first days of hydration, the formation of minor quantities of portlandite and monocarbonate is calculated. The hydration of the Portland cement, which produces a surplus of portlandite, is initially relatively fast, while the dissolution of the silica fume, which consumes $Ca(OH)_2$ to precipitate C-S-H, is somewhat slower but continues more steadily. The temporary occurrence of portlandite and monocarbonate during the first few days agrees with the experimental observations in Fig. 4.2 and emphasizes that not only the composition of solids but also their reactivity determines the phase assemblage.

The thermodynamic modeling of LAC predicts similarly C-A-S-H with Ca/Si of 1 as the main hydration product. In addition, ettringite, hydrotalcite, and small quantities of strätlingite were predicted, which agrees well with the experimental findings [10].

4 Influence of Increasing SiO_2 Content

The main difference between the two blended cements and the ordinary Portland cement was the complete absence of portlandite (Fig. 4.4) and the presence of a C-S-H with a lower Ca/Si and a higher Al/Si ratio. The presence of a low Ca/Si C-S-H led also to an enhanced uptake of alkalis and to lower pH values in the blended cements. While the pH of the OPC mixture was after 3.5 years of hydration time at 13.3, it was at 12.3 for the LAC cement and as low as 11.5 for the ESDRED cement (Fig. 4.1). Ettringite has been found in all cements investigated (Fig. 4.4).

Thermodynamic modeling may also be used to calculate changes in the phase assemblage and in the pore solution caused by the blending of Portland cement with SiO_2 (Fig. 4.5). The addition of moderate amounts of SiO_2 leads to the consumption of portlandite, to the formation of more C-S-H, and eventually to a lowering of the Ca/Si ratio. Increasing amounts of low Ca/Si C-S-H lead to an increased fraction of the total Al taken up in C-S-H. If more SiO_2 is available, the destabilization of monocarbonate is calculated. Finally, ettringite becomes unstable with regard to $Al(OH)_3$ and gypsum is calculated to form due to the decrease of the pH value below 10. However, more probably the sulfate will be taken up in the C-S-H [1], thus preventing gypsum formation. Sulfate uptake by C-S-H was not considered in the thermodynamic model used.

The composition of the OPC studied after 1 year refers to composition at the left side of Fig. 4.5, while ESDRED and LAC after 1 year of hydration time will be around 25 and 28 % replacement level, respectively, as only a part of the silica fume and the slag had reacted. With more time, as the silica fume and also the unhydrated slag continue to react slowly, the mixtures studied will move more towards the right side of Fig. 4.5.

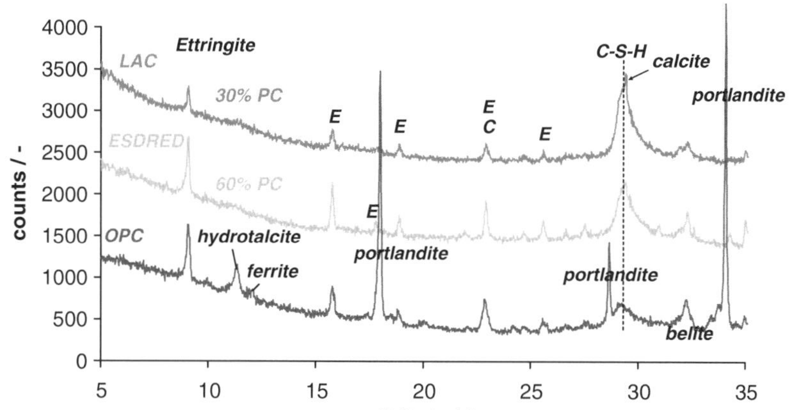

Fig. 4.4 XRD of an OPC and two blended cements: LAC (blended with 10 % SiO_2) and the shotcrete mixture ESDRED (blended with 40 % SiO_2), hydrated for 3.5 years. *C* calcite, *E* ettringite

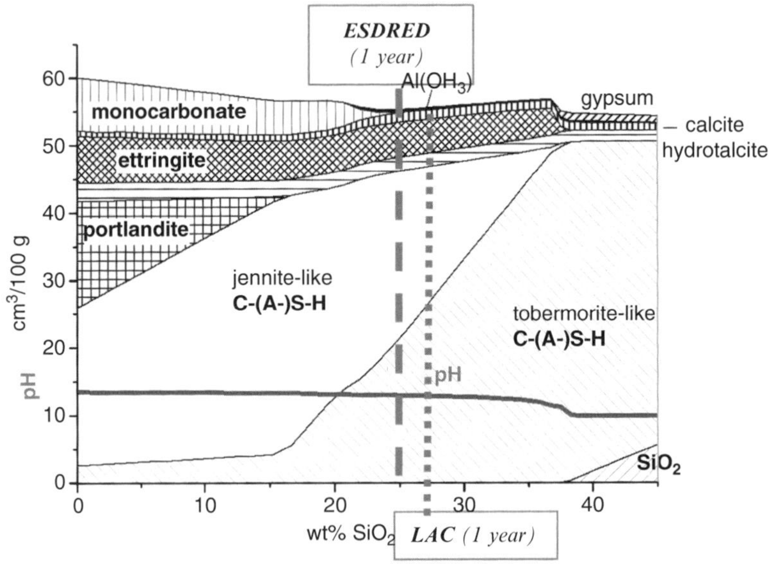

Fig. 4.5 Modeled changes in hydrated Portland cement upon blending with SiO_2. Adapted from Lothenbach et al. (2011); reprinted with permission

5 Conclusions

The composition of the solid phases and the pore solutions of the three cements studied were clearly different after 1 year of hydration. The OPC continued to have significant amounts of portlandite and high pH values in the pore solution. The presence or absence of portlandite affected the composition of C-S-H; in OPC a C-S-H with a high Ca/Si was present, while LAC and ESDRED had a C-S-H with a lower Ca/Si ratio of approx. 1.0. The presence of a lower Ca/Si C-S-H also led to an enhanced uptake of Al and alkalis by C-S-H. Ettringite was observed in all three mixes, with ESDRED and OPC containing significantly more ettringite than LAC. The composition of the pore solution after 3.5 years differed widely. The ESDRED that contained most SiO_2 showed with 11.5 the lowest pH values. It is possible that these values could drop even further as significant quantities of silica fume had not yet reacted.

Thermodynamic modeling indicates that the replacement of PC by SiO_2 leads to very significant changes of the composition of the hydrate assemblage. Not only portlandite is depleted to form more C-S-H but also the amount and kind of AFm and AFt phases are affected. Monocarbonate is calculated to be unstable with regard to strätlingite and calcite as more and more SiO_2 is available. If even higher quantities of SiO_2 are available, eventually ettringite is calculated to be unstable as the pH decreases.

Acknowledgments The partial financial support of the Cement Clay Interaction (CI) experiment at Mont Terri is gratefully acknowledged. Thanks to G. Le Saout (Empa), M. Ben Haha, E. Wieland (PSI), and B. Schwyn (nagra) for helpful discussions.

References

1. Barbarulo R, Peycelon H, Prene S (2003) Experimental study and modelling of sulfate sorption on calcium silicate hydrate. Annales de Chimie: Science des materiaux 1:S5–S10
2. Hong S-Y, Glasser FP (1999) Alkali binding in cement pastes. Part I. The C-S-H phase. Cement Concr Res 29:1893–1903
3. Hummel W, Berner U, Curti E, Pearson FJ, Thoenen T (2002) Nagra/PSI Chemical Thermodynamic Data Base 01/01. Universal Publishers/uPUBLISH.com
4. Kulik D, Berner U, Curti E (2004) Modelling geochemcial equilibrium partitioning with the GEMS-PSI Code. PSI Scientific Report 2003 IV, pp 109–121
5. Larbi JA, Fraay ALA, Bijen JM (1990) The chemistry of the pore fluid of silica fume-blended cement systems. Cement Concr Res 20:506–516
6. Le Saout G, Kocaba V, Scrivener KL (2011) Application of the Rietveld method to the analysis of anhydrous cement. Cement Concr Res 41:133–148
7. Lothenbach B, Matschei T, Möschner G, Glasser FP (2008) Thermodynamic modelling of the effect of temperature on the hydration and porosity of Portland cement. Cement Concr Res 38:1–8
8. Lothenbach B, Le Saout G, Gallucci E, Scrivener K (2008b) Influence of limestone on the hydration of Portland cements. Cement and Concrete Research 38:1162–1186
9. Lothenbach B, Scrivener K, Hooton RD (2011) Supplementary cementitious materials. Cement and Concrete Research 41:1244–1256
10. Lothenbach B, Le Saout G, Ben Haha M, Figi R, Wieland E (2012) Hydration of low-alkali CEM III/B -SiO2 blend (LAC). Cement and Concrete Research, 42:410–423
11. Parrot LJ, Killoh DC (1984) Prediction of cement hydration. British Ceramic Proceedings 35:41–53

Chapter 5
Application of a Sub-lattice Model to Predictions of Cement Hydrate Chemistry

J.A. Gisby, M. Tyrer, R.H. Davies, A.T. Dinsdale, C.S. Walker, and F.P. Glasser

1 Introduction

Applications of computational thermodynamics to the field of cement chemistry have seen considerable advances in recent years, owing both to developments in numerical methods and the increased availability of fundamental data. These advances have added considerably to our understanding of the evolution of nuclear waste repository chemistry, especially in the prediction of groundwater–waste–cement–rock interactions in and around the near field.

The engineered component of a low- and intermediate-level nuclear waste repository in the UK will almost certainly consist primarily of ordinary Portland cement (OPC)-based materials. OPC provides a chemical barrier by its ability to buffer the pH > 12 as it dissolves in a percolating groundwater. A high pH in a repository is desirable because it helps to minimise the solubility of many radionuclides, metal corrosion and microbial activity. In order to assess the likely performance of the chemical barrier, reliable models are required with which to make predictions beyond the spatial and temporal limits imposed by experiment and observation. The application of thermodynamic modelling to cement chemistry has been advocated by numerous workers over the last two decades [1–5]. Its development has been driven largely by the need for a predictive capability in

J.A. Gisby • R.H. Davies • A.T. Dinsdale
National Physical Laboratory, Hampton Road, Teddington TW11 0LW, UK

M. Tyrer (✉)
Mineral Industry Research Organisation, Concorde House, Solihull, Birmingham B37 7UQ, UK
e-mail: m.tyrer@miro.co.uk

C.S. Walker
Japan Atomic Energy Agency (JAEA), 4-33 Muramatsu, Tokai, Naka, Ibaraki 319-1194, Japan

F.P. Glasser
Department of Chemistry, University of Aberdeen, Meston Walk, Old Aberdeen AB24 3FX, UK

F. Bart et al. (eds.), *Cement-Based Materials for Nuclear Waste Storage*,

DOI 10.1007/978-1-4614-3445-0_5,

modelling the near-field processes, which govern the performance of proposed deep geological disposal repositories for nuclear waste. Early work [1–5] sought to develop pragmatic models with which to simulate the thermodynamic evolution of cements in the repository environment. In achieving this goal, of critical importance was the development of a robust description of the incongruent dissolution of C-S-H gel because of its major contribution to the longevity of the chemical barrier.

It is expected that a nuclear waste repository would experience elevated temperatures over time scales of thousands of years, either due to the ambient temperature of the host rock (e.g., 40–45 °C at 1 km depth in Japan), the heat of hydration of the cement or radiolytic heating. Long-term experiments at elevated temperatures have been shown to have a marked effect on the solubility behaviour of C-S-H gel [6]. Thus, developing a credible predictive model, which can be used to describe the dissolution of C-S-H gel at room and elevated temperatures, is of critical importance.

2 Previous Work

The challenge of representing the incongruous dissolution of C-S-H gel by a numerical model has been met by numerous workers using several different approaches. Space precludes a detailed appraisal of them all, but the methods have been described in recent reviews of this field [6–8]. One approach is to develop an empirical (or semi-empirical) model based on solid-solution and aqueous-solution compositions measured by experiment. In these cases, either one or two pure solid phases with variable solubility products are chosen and the model parameters are arbitrarily adjusted until a match is made with the measured solubility data of the C-S-H system [1–4]. Whilst these models are pragmatic and effective, they often do not account for all Ca/Si ratios (0.8–2.7) of the C-S-H system expected in hydrated/leached OPC (Fig. 5.1), do not have a firm basis in thermodynamic theory or do not allow simple expansion to include elevated temperatures or variations in solid phase composition. Despite these observations, the Berner [3] model and its variants have been the most widely used owing to its elegant simplicity. A more thermodynamically rigorous description of the C-S-H system uses a solid-solution aqueous-solution-based model [9–11, 35], which has only been applied to Ca/Si ratios greater than 1.0 in the C-S-H gel (Fig. 5.1) and at room temperatures. True Gibbs energy-based models (of which this chapter reports one example) strive for greater flexibility and rely more on thermodynamic realism than earlier methods. They can therefore account for all Ca/Si ratios expected in the C-S-H gel (Fig. 5.1) and extend the limits of temperature.

Vertical dashed lines at Ca/Si ratios = 0.8, 1.7 and 2.7 are for visual reference.

Model references are 87Atk [1], 87Gla [2], 88Ber [3], 92Rea [4], 96Ker [10], 07Wal [7], 97Bör [11], 99Rah [12], 98Tho [24], 07Gis, current model, solved in 2D.

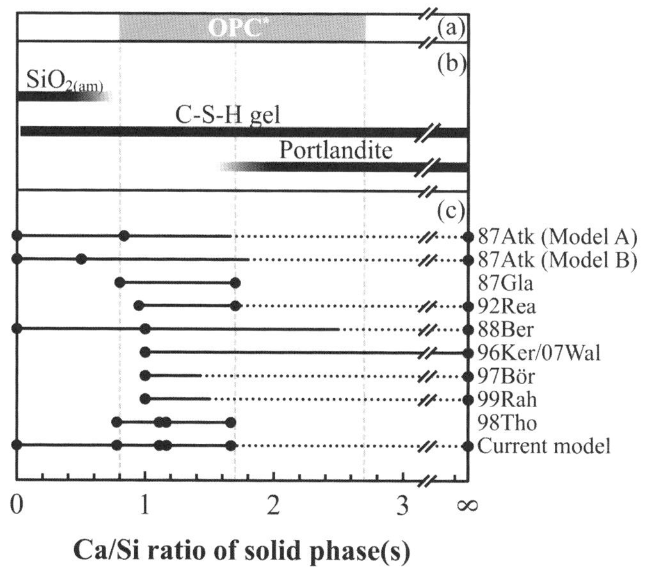

Fig. 5.1 Schematic representation of Ca/Si ratio of C-S-H solid phase(s): (**a**) expected in *pure water hydrated/leached OPC, (**b**) based on studies of synthetic C-S-H gels, where silica gel or portlandite may co-exist with the gel phase, (**c**) in C-S-H gel models, *Horizontal solid lines* represent compositional ranges of variable solubility and *dotted lines* of fixed solubility. Compositions that form an integral part of each model are shown as circles. Note that although several models operate numerically over a wide range of Ca:Si ratios in the solid phase, this is a numerical consequence of the modelling approach. The authors do not imply that the CSH phase shows stability beyond the limits determined by experiment, so no such inference should be drawn

3 Modelling Approach

The method adopted here uses a compound energy model [13] applied using a sub-lattice representation of the C-S-H gel guided by structural studies of this phase. To implement these ideas, thc codc MTDATA [14] has been used throughout. This is a reliable and general software tool for calculating phase and chemical equilibria involving multiple solution or stoichiometric phases with ease and reliability. It provides true Gibbs energy minimisation through the solution of a non-linear optimisation problem with linear constraints using the National Physical Laboratory (NPL) Numerical Optimisation Software Library (NOSL), which guarantees mathematically that the Gibbs energy reduces each time it is evaluated. MTDATA has a long history of application dating back to the 1970s in fields such as ferrous and non-ferrous metallurgy [15–17], slag and matte chemistry [18–20], nuclear accident simulation [21], molten salt chemistry [22] and cement clinkering [23]. The CALPHAD modelling principle embodied within MTDATA [14] is that a database developer derives parameters to accurately represent the Gibbs energy of each phase that might form in a system as a function of temperature, composition and, if necessary, pressure. This is achieved using data assessment tools within MTDATA, which allow parameters to be optimised to give the best possible agreement between calculated thermodynamic properties (such as heat capacities, activities, enthalpies of mixing), phase equilibria (solubilities) and collated experimental properties. The aim of the work described here has been to develop a flexible, yet rigorous model and associated thermodynamic data for the C-S-H system, at room and elevated temperatures, which are compatible with existing NPL oxide and aqueous species databases.

In the current model, the C-S-H system was represented by a SiO_2 gel (Ca/Si ratio = 0–1.11), C-S-H gel (Ca/Si = 0.78–1.67) and portlandite present as a distinct phase at higher Ca/Si ratios. The C-S-H gel was represented using five sub-lattices, as in [24], with occupancies as follows:

Sub-lattice	Occupancy	Number and type of sites
1	CaO	1/3 interlayer site, 2 sites pairing with SiO_2
2	CaO, H_2O	1 interlayer site
3	SiO_2	2 sites pairing with CaO
4	SiO_2, Va	1 site representing bridging SiO_2 or Vacancy
5	H_2O	6 sites

Unaries formed by taking one species from each sub-lattice in turn, i.e.

$$(\mathrm{CaO})_{7/3}(\mathrm{CaO})_1(\mathrm{SiO_2})_2(\mathrm{H_2O})_1(\mathrm{H_2O})_5,$$
$$(\mathrm{CaO})_{7/3}(\mathrm{H_2O})_1(\mathrm{SiO_2})_2(\mathrm{H_2SiO_3})_1(\mathrm{H_2O})_5,$$
$$(\mathrm{CaO})_{7/3}(\mathrm{CaO})_1(\mathrm{SiO_2})_2(\mathrm{H_2SiO_3})_1(\mathrm{H_2O})_5 \text{ and}$$
$$(\mathrm{CaO})_{7/3}(\mathrm{H_2O})_1(\mathrm{SiO_2})_2(\mathrm{H_2O})_1(\mathrm{H_2O})_5,$$

were assigned Gibbs energies varying with temperature according to a standard function of the form:

$$A + B\mathrm{T} + C\mathrm{TlnT} + D\mathrm{T}^2 + E\mathrm{T}^3 + F/\mathrm{T}$$

The *C*, *D*, *E* and *F* parameters were obtained from estimates of heat capacity, Cp(T), for each species guided by those for compounds like portlandite ($Ca(OH)_2$) and plombierite (14 Å tobermorite), for which thermodynamic data were available from the SGTE substance database [25]. Initial values for the entropy-like *B* parameters and enthalpy-like *A* parameters were estimated on a similar basis. The *A* parameters were then adjusted, along with others introduced to represent excess Gibbs energies of interaction between sub-lattice species, using MTDATA's data assessment tools to reproduce measured compositions of C-S-H gel and aqueous phases in equilibrium from data available in the literature [1, 26–35].

The SiO_2 gel was represented using two sub-lattices with occupancies as follows (where Va is used to indicate a potentially vacant sub-lattice):

Sub-lattice	Occupancy	Number of sites
1	$Ca_{7/3}H_{14}O_{28/3}$, $Ca_{10/3}H_{12}O_{28/3}$, Va	1
2	SiO_2	3

This model allows the composition of the SiO_2 gel phase to vary within a triangle formed by the unaries $(SiO_2)_3$, $(Ca_{7/3}H_{14}O_{28/3})_1(SiO_2)_3$ and $(Ca_{10/3}H_{12}O_{28/3})_1(SiO_2)_3$ as opposed to a line used by Kulik and Kersten [5]. The first two of these unaries lie along the SiO_2-rich side of the composition parallelogram formed by the four C-S-H gel unaries. Data for the pure SiO_2 unary were based upon those for amorphous SiO_2 in the SGTE substance database.

Parameters representing interactions between species on the first sub-lattice in this phase were derived, as for the C-S-H gel phase, to represent experimental phase equilibrium and experimental solubility data as closely as possible. Calculations are most flexibly undertaken in MTDATA by specifying start and end compositions at a fixed temperature or start and end temperatures with a fixed composition. The program will perform a series of phase equilibrium calculations for conditions varying between the specified extremes and results can be plotted in terms of phase amounts, aqueous species, element distributions or pH values. Thermodynamic equilibria between C-S-H and SiO_2 gels, portlandite and an aqueous phase could, for example, be calculated by stepping across the H_2O-SiO_2-$Ca(OH)_2$ composition triangle between two fixed compositions. In Figs. 5.2, 5.3 and 5.4, a system comprising 1 kg H_2O and 1 mole of SiO_2 is simulated, whilst stepping the CaO content from 0 to 2.5 moles.

Since the data for the C-S-H and SiO_2 gels, for other stoichiometric phases such as quartz and portlandite and for aqueous species (in the MTDATA implementation of the SLOP98 Database) all have Gibbs energies modelled as a function of temperature, as already described, equilibria can be calculated at elevated temperatures just as at room temperature using these data.

4 Results

4.1 C-S-H Chemistry

Figures 5.2, 5.3 and 5.4 show the predicted pH and calcium and silicon concentrations in solution at equilibrium with the SiO_2 gel, C-S-H gel and portlandite, as constrained by the Ca/Si ratio of the system, at 25 °C and the corresponding measured values reported in the literature at near room temperature.

It should be noted that such gels are generally considered to be metastable because they continue to react very slowly, eventually forming crystalline CaO–SiO_2–H_2O phases. MTDATA predicts this true equilibrium state but metastable equilibria can be

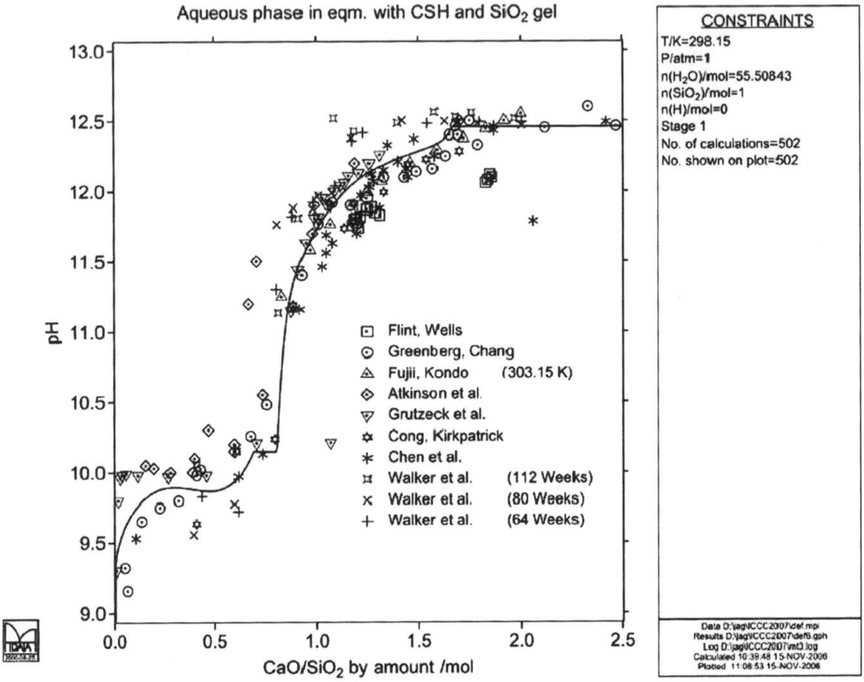

Fig. 5.2 Predicted pH of solution at equilibrium with C-S-H compared with measured values reported in the literature

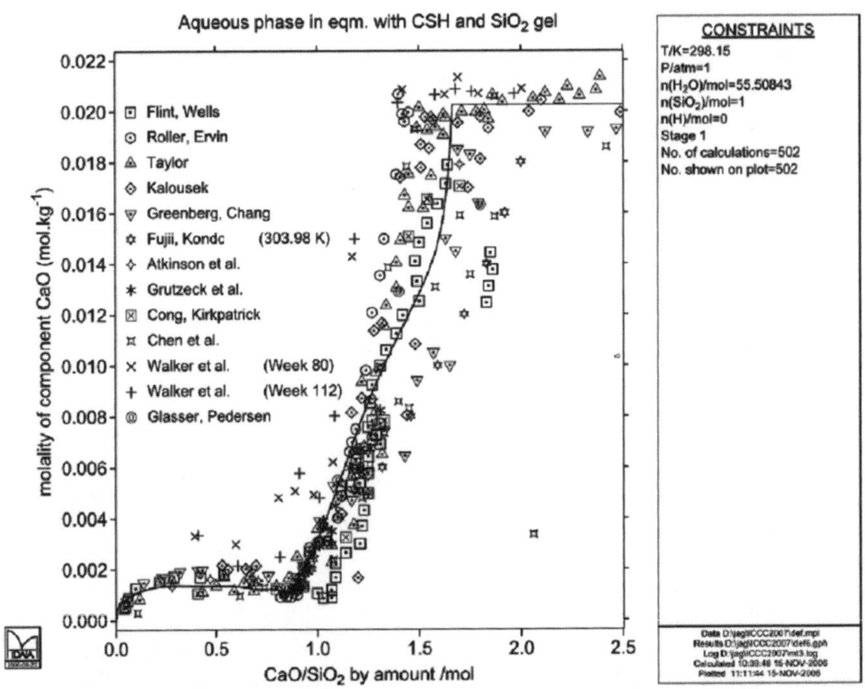

Fig. 5.3 Predicted calcium concentration of a solution at equilibrium with C-S-H compared with measured values reported in the literature

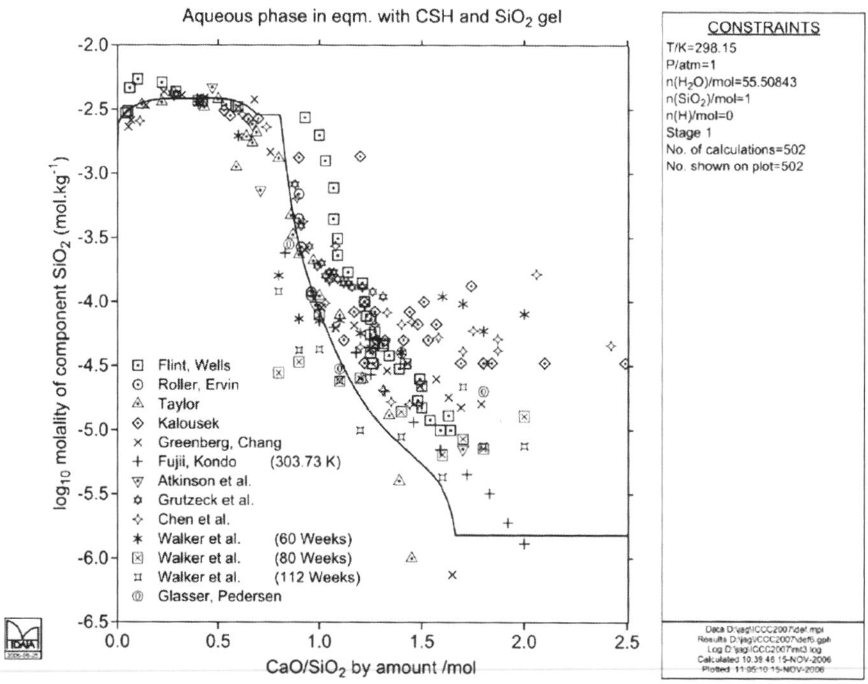

Fig. 5.4 Predicted silicon concentration of a solution at equilibrium with C-S-H compared with measured values reported in the literature

studied by "classifying as absent" long-term decomposition products or removing them from consideration in the calculations.

It has been noted by several authors that the experimental data divide into two populations, depending on whether the C-S-H gel has been prepared directly from ions in solution (i.e., by direct reaction or double decomposition) or from the hydration and subsequent leaching of tricalcium silicate (the latter route forming the least soluble solid). For the purposes of this model, parameters have been derived to reproduce the lower SiO_2 solubility curve as described by Jennings [41], thought to be appropriate to the most stable gel structure, although reproducing either curve is relatively straightforward. It should be noted that the high pH plateau is a feature of the co-existence or portlandite and the C-S-H gel at Ca/Si ratios $\geq$1.7.

4.2 Elevated Temperatures

Surprisingly few long-term experiments of C-S-H gels maintained at high temperatures have been reported. Work by Glasser et al. [6] has been chosen with which to compare the results of the predictive calculations as shown in Fig. 5.5.

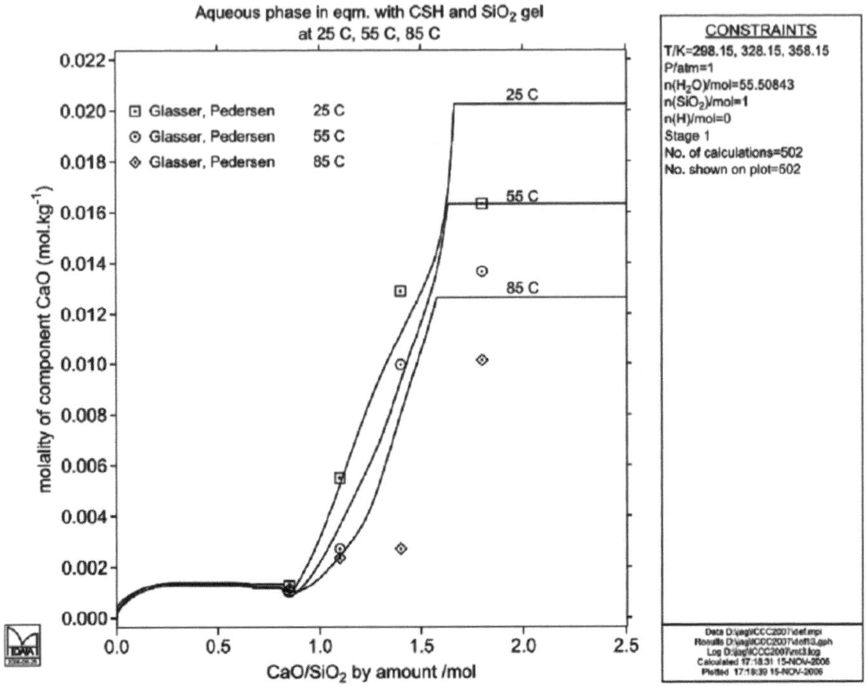

Fig. 5.5 Predicted calcium concentration of solutions at elevated temperature after equilibrium with C-S-H compared with experimentally measured values

It should be noted that the model predictions were not fitted to the experimental data shown. The order and shape of each curve in Fig. 5.5 is similar for both predicted and observed results, reflecting a decreasing solubility of C-S-H gel with increasing temperature. The general magnitude of the changes in solubility with temperature is predicted well, although the match between absolute solubility values at high Ca/Si ratios appears poor. When Fig. 5.3 is taken into account, however, it is clear that the room temperature solubility results of [6] for Ca/Si ratios >1.5 are not representative of the bulk of the experimental solubility data available. As more high temperature experimental data become available there is every reason to believe that the current calculated results will provide a better match or, should this not be the case, further optimisation of the Gibbs energies adopted would be possible.

4.3 *Leaching*

Simulation of leaching has been attempted in the simplest possible way by repeatedly replacing the aqueous phase of the C-S-H–water system, whilst performing a new equilibrium calculation at each stage. In this way, the incongruous dissolution of the

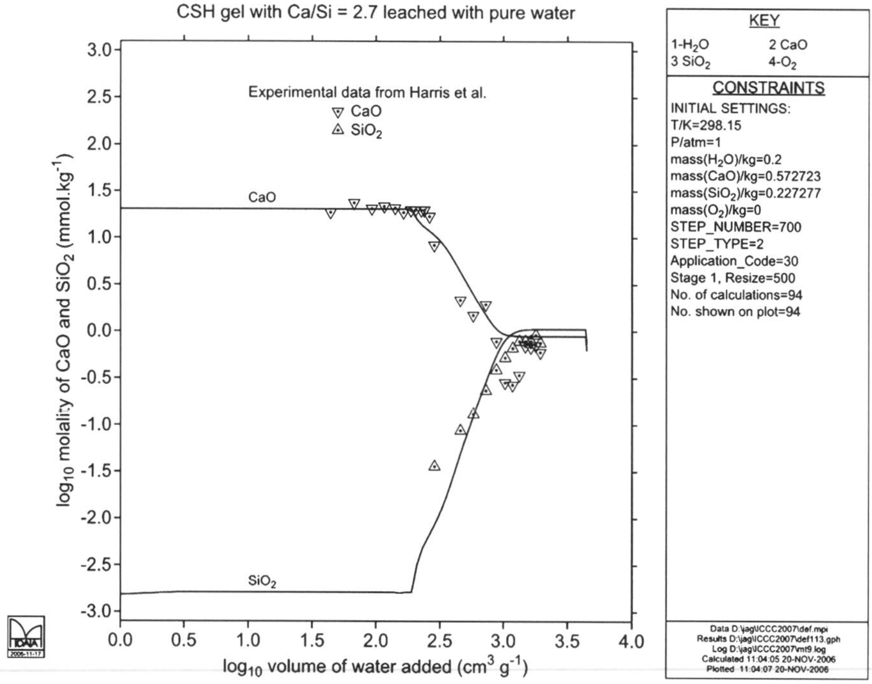

Fig. 5.6 Concentration of calcium and silicon in solution during the simulated leaching of Portlandite–C-S-H mixtures in comparison with experimental modelling of this system by Harris et al. [9]. At Ca/Si $\geq$ 1.7 portlandite is expected to be present with the C-S-H gel

C-S-H gel is demonstrated, as shown in Fig. 5.6. Significantly more advanced simulations are possible, e.g. to account for changes in dissolution, porosity and flow, through coupling MTDATA to more realistic transport codes.

4.4 Environmentally Sensitive Applications

Arsenic was chosen as an example of an element of environmental concern which is both amphoteric and redox sensitive. A common component of wastes from mineral processing, it also binds to C-S-H gel in addition to replace sulphate in ettringite, but experiments suggest that the majority of this element is likely to be incorporated into cements as calcium arsenic compounds [36], both in the form of As(III) and As(V).

In this study of arsenic chemistry, a simplified representation of the As(III) aqueous chemistry has been adopted using a neutral $HAsO_2$ (equivalent to $As(OH)_3$–H_2O) species and the single dissociated ion $AsO(OH)^{2-}$. For As(V) a complete series of

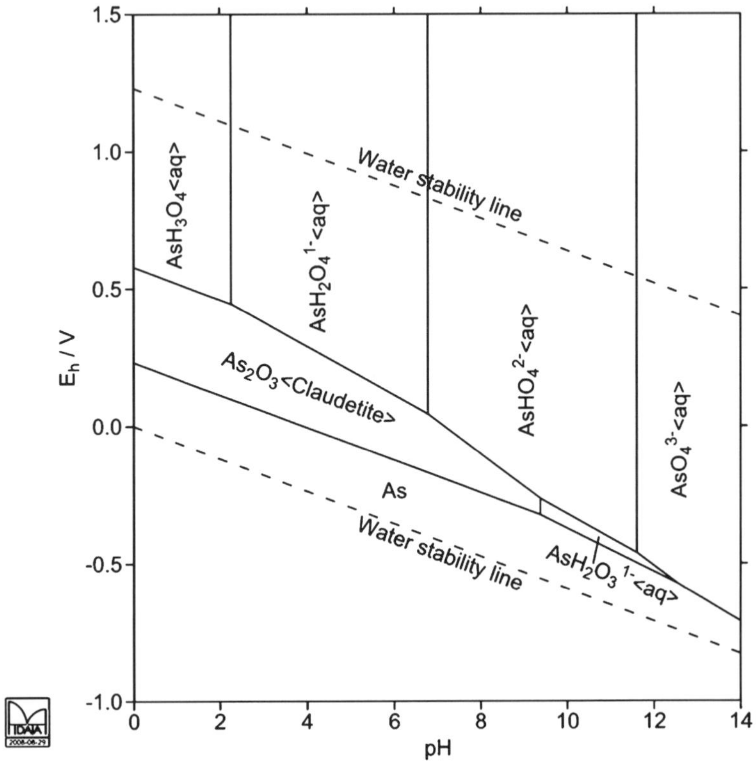

Fig. 5.7 Calculated Pourbaix diagram for As-containing species at 25 °C

hydrolysis products from AsO_4^{3-} to $AsO(OH)_3$ are included. Figure 5.7 is a calculated Pourbaix diagram for As at 25 °C, showing the extent of As_2O_3 (claudetite) stability and the distribution of aqueous As(III) and As(V) species as a function of oxidation potential (Eh) and pH. Dashed lines indicate infeasible regions beyond which the aqueous solution breaks down through the formation of gaseous species.

The temperature dependence of the Gibbs energy of all species developed in this work was modelled to be fully consistent with reference states used throughout the worldwide modelling group SGTE [25]. Thermodynamic data for As_2O_3 (Claudetite) from the SGTE substance database were adjusted slightly to improve consistency with HATCHES [37] at room temperature and experimental measurements [38] from 0 to 75 °C.

Data for $CaAs_2O_4$ were taken from the SGTE substance database [25] without change. Initial data for $CaAsO_2(OH)$ were based upon a combination of those for $CaAs_2O_4$ and portlandite with an estimated enthalpy of formation, adjusted within likely limits of experimental measurement, to give reasonable calculated Ca^{2+} and As^{3+} solubilities as a function of temperature and pH.

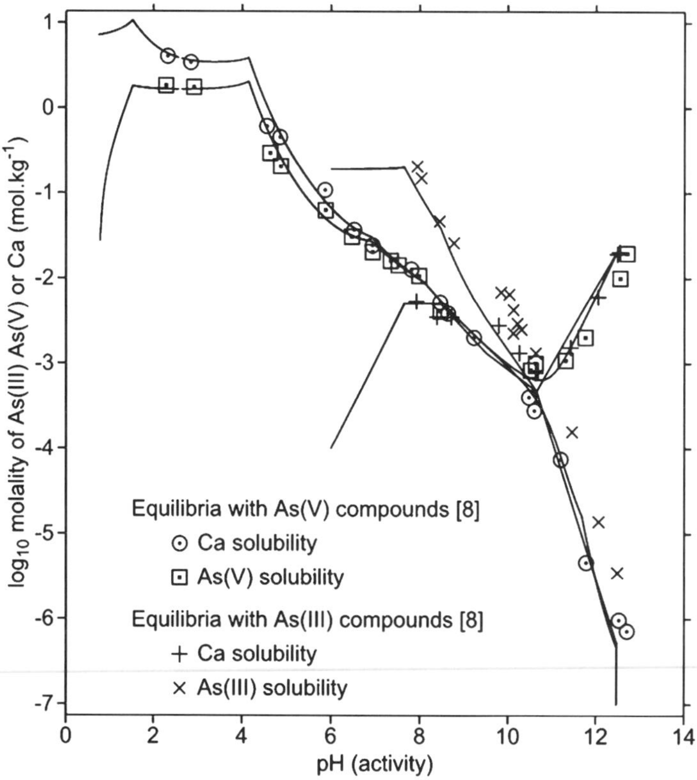

Fig. 5.8 Calculated Ca and As solubilities in solutions equilibrated with As(III)- and As(V)-containing compounds in turn compared with data from [39]

According to measurements from [38], the H_2O–As_2O_5 system features two hydrated arsenic compounds in the temperature range 0–100 °C, namely $As_2O_5 \cdot 4H_2O$ and $3As_2O_5 \cdot 5H_2O$. Thermodynamic data for these species were derived so as to reproduce their reported solubilities in water [38] over this complete temperature range. Data for the calcium arsenate hydrates, $Ca_2AsO_4(OH)$, $CaAsO_3(OH)$, $CaAs_2O_4(OH)_4$, and $Ca_5As_4O_{14}(OH)_2$ were derived based upon combinations of those for simpler species with an initially estimated, then adjusted enthalpy of formation, in a similar way to that outlined for $CaAsO_2(OH)$. Data for $Ca_3As_2O_8$ were taken from the SGTE substance database [25] without change.

Calculated Ca, As(III) and As(V) solubilities in equilibrium with As(III) and As(V) compounds, expressed as a function of pH, show excellent agreement with reported values [39] (Fig. 5.8). Calculated As(III) and As(V) solubilities in equilibrium with As(III) and As(V), expressed as a function of Ca concentration, are consistent with experimental measurements from [38, 40].

5 Discussion

The current chapter illustrates the potential of using a compound energy model to represent C-S-H gel when simulating the chemistry of cements. We have shown the basis on which the sub-lattice model is built and demonstrated its application in simulating simple equilibrium chemistry at low and elevated temperatures, incongruous dissolution during leaching and as part of a more complex model of arsenic speciation in a cemented waste form. Development of this approach is the subject of a new study at NPL, in which solid solution–aqueous solution models will require a more complex representative C-S-H gel chemistry. For example, incorporation of uranium, aluminium, sulphate and halides all pose problems which can readily be addressed by considering their likely fate in the solid phase.

References

1. Atkinson A, Hearne JA, Knights CF (1987) Aqueous and thermodynamic modeling of CaO-SiO_2-H_2O gels. UK Atomic Energy Authority, Report AERER12548
2. Glasser FP, MacPhee DE, Lachowski EE (1987) Solubility modeling of cements: implications for radioactive waste immobilization. Mat Res Soc Symp Proc 84:331–341
3. Berner UR (1988) Modeling the incongruent dissolution of hydrated cement minerals. Radiochim Acta 44(45):387–393
4. Reardon EJ (1992) Problems and approaches to the prediction of the chemical composition of cement/water systems. Waste Manage 12:221–239
5. Kulik DA, Kersten M (2001) Aqueous solubility diagrams for cementitious waste stabilization systems: II, End-member stoichiometries of ideal calcium silicate hydrate solid solutions. J Am Ceram Soc 84:3017–26
6. Glasser FP, Pedersen J, Goldthorpe K, Atkins M (2005) Solubility reactions of cement components with NaCl solutions: I. Ca(OH)2 and C-S-H. Adv Cem Res 17:57–64
7. Benbow S, Walker C, Savage D (2007) Intercomparison of cement solid-solution models. Issues affecting the geochemical evolution of repositories for radioactive waste. SKI Report 2007:29 ISSN 1104 1374
8. Soler JM (2007) Thermodynamic description of the solubility of C-S-H gels in hydrated Portland cement. Literature review. POSIVA Working Report 2007–88
9. Harris AW, Manning MC, Tearle WM, Tweed CJ (2002) Testing of models of the dissolution of cements—leaching of synthetic CSH gels. Cement Concr Res 32:731–746
10. Kersten M (1996) Aqueous solubility diagrams for cementitious waste stabilization systems. 1. The C-S-H solid-solution system. Environ Sci Technol 30:2286–2293
11. Börjesson S, Emrén A, Ekberg C (1997) A thermodynamic model for the calcium silicate hydrate gel, modelled as a non-ideal binary solid solution. Cement Concr Res 27:1649–1657
12. Rahman MM, Nagasaki S, Tanaka S (1999) A model for dissolution of CaO-SiO_2–H_2O gel at $Ca/Si > 1$. Cement Concr Res 29:1091–1097
13. Barry TI, Dinsdale AT, Gisby JA et al (1992) The compound energy model for ionic solutions with applications to solid oxides. J Phase Equilibria 13:459–475
14. Davies RH, Dinsdale AT, Gisby JA et al (2002) MTDATA: Thermodynamics and phase equilibrium software from the National Physical Laboratory. CALPHAD 26:229–271
15. Hunt C, Nottay J, Brewin A, Dinsdale AT (2002) NPL Report MATC(A) 83
16. Wang J, van der Zwaag S (2001) Composition design of a novel P containing TRIP steel. Z Metallkd 92:1299–1311

17. Putman DC, Thomson RC (2003) Modeling microstructural evolution of austempered ductile iron. Int J Cast Metals Res 16:191–196
18. Taskinen P, Dinsdale AT, Gisby JA (2004), Industrial slag chemistry—a case study of computational thermodynamics. Metal Separation Technologies III, Copper Mountain, Colorado
19. Gisby JA, Dinsdale AT, Barton-Jones I et al (2002), Phase equilibria in oxide and sulphide systems, in sulfide smelting. In: Stephens RL, Sohn HY (eds) TMS 2002 131st Annual Meeting & Exhibition, Seattle
20. Barry TI, Dinsdale AT, Gisby JA (1993) Predictive thermochemistry and phase equilibria of slags. JOM 45:32–38
21. Ball RGJ, Mignanelli MA, Barry TI, Gisby JA (1993) The calculation of phase equilibria of oxide core-concrete systems. J Nucl Mater 201:238–249
22. Barry TI, Dinsdale AT (1987) Thermodynamics of metal-gas-liquid reactions. Mater Sci Technol 3:501–511
23. Barry TI, Glasser FP (2000) Calculation of Portland cement clinkering reactions. Adv Cem Res 12:19–28
24. Thomas JJ, Jennings HM (1998) Free-energy-based model of chemical equilibria in the CaO-SiO_2-H_2O system. J Am Ceram Soc 83:606–612
25. SGTE substance database, Thermodynamic properties of inorganic materials. Landolt-Börnstein Group IV (Physical Chemistry), vol 19, Springer, Berlin
26. Flint EP, Wells LS (1934) Study of the system CaO-SiO_2–H_2O at 30°C and of the reaction of water on anhydrous calcium silicates. J Res Natl Bur Stan 12:751–783
27. Roller PS, Ervin G Jr (1940) The system calcium-silica-water at 30°. The association of silicate ion in dilute alkaline solution. J Am Chem Soc 62:461–471
28. Taylor HFW (1950) Hydrated calcium silicates. Part I. Compound formation at ordinary temperature. J Chem Soc 726:3682–3690
29. Kalousek G (1952) Application of differential thermal analysis in a study of the system lime-silica-water. In: Third international symposium on the chemistry of cement, Cement and Concrete Association, London
30. Greenberg SA, Chang TN (1965) Investigation of the colloidal hydrated calcium silicates. II. Solubility relationships in the calcium-silica-water system at 25°C. J Phys Chem 69:182–188
31. Fujii K, Kondo W (1981) Heterogeneous equilibrium of calcium silicate hydrate in water at 30°C. J Chem Soc/Dalt Trans 2:645–651
32. Grutzeck M, Benesi A, Fanning B (1989) Silicon-29 magic angle spinning nuclear magnetic resonance study of calcium silicate hydrates. J Am Ceram Soc 72:665–668
33. Cong X, Kirkpatrick RJ (1996) 29Si MAS NMR study of the structure of calcium silicate hydrate. Adv Cem Bas Mat 3:144–156
34. Chen JJ, Thomas JJ, Taylor HFW, Jennings HM (2004) Solubility and structure of calcium silicate hydrate. Cement Concr Res 34:1499–1519
35. Walker CS, Savage D, Tyrer M, Ragnarsdottir KV (2007) Non-ideal solid solution aqueous solution modeling of synthetic calcium silicate hydrate. Cement Concr Res 37:502–511
36. Phenrat T, Marhaba TF, Rachakornkij M (2005) J Hazard Mater 118(1–3):185–195
37. Bond KA, Heath TG, Tweed CJ, MTDATA implementation of HATCHES: A referenced thermodynamic database for chemical equilibrium studies (1997) Nirex NSS/R379
38. Linke WF (1958) Solubilities of inorganic and metal-organic compounds (Ar-In), 4th edn., vol. 1, D. Van Nostrand Company Inc., Princeton, New Jersey
39. Nishimura T, Itoh CT, Tozawa K (1987) Arsenic Metallurgy, Fundamentals and Applications, pp 77–89
40. Stronach SA, Walker NL, Macphee DE, Glasser FP (1997) Reactions between cement and As(III) oxide: the system CaO-SiO_2-As_2O_3-H_2O at 25°C. Waste Manag 17:9–13
41. Jennings HM (1986) Aqueous solubility relationships for two types of calcium silicate hydrate. J Am Ceram Soc 69:614–618

Chapter 6
Characterization and Solubility Determination of the Solid-Solution Between AFm-I_2 and AFm-SO_4

L. Aimoz, E. Wieland, D.A. Kulik, B. Lothenbach, M.A. Glaus, and E. Curti

1 Introduction

Radioactive waste contains significant amounts of ^{129}I. ^{129}I will be released as dissolved anionic species over time, since iodide ($I^{\#}$) is the chemically stable form of iodine under repository conditions. ^{129}I is predicted to be dose-determining in safety analyses due to its long half-life (1.57×10^7 years) and its weak retention by the major minerals from the engineered barriers of underground disposal facilities and surrounding host rocks [1]. Such minerals are negatively charged (especially clay minerals), and, therefore, only weakly retard anionic species. An important unresolved issue is the identification of specific solid phases that would preferentially bind iodide in cement-stabilized radioactive waste. A previous study showed that calcium silicate hydrate (C-S-H) phase, which is the most abundant component of hydrated cement, is not the main scavenger of iodide [2]. Minerals with a positive structural charge, in particular AFm phases, were suggested to be candidates for iodide retention in the cement matrix [2–5].

AFm-type cement minerals are present in hydrated Portland cement and belong to the family of layered double hydroxides (LDH). AFm phases ($[Ca_2Al(OH)_6]$

L. Aimoz (✉)
Laboratory for Waste Management (LES), Paul Scherrer Institute (PSI), Villigen, Switzerland

Institute of Geological Sciences, University of Bern, Bern, Switzerland
e-mail: Laure.Aimoz@psi.ch

E. Wieland • D.A. Kulik • M.A. Glaus • E. Curti
Laboratory for Waste Management (LES), Paul Scherrer Institute (PSI), Villigen, Switzerland

B. Lothenbach
Empa, Lab. Concrete & Construction Chemistry, Dübendorf, Switzerland

Swiss Federal Laboratories for Materials Science and Technology (EMPA), Concrete Construction Chemistry Laboratory, Dübendorf, Switzerland

F. Bart et al. (eds.), *Cement-Based Materials for Nuclear Waste Storage*,
DOI 10.1007/978-1-4614-3445-0_6, © Springer Science+Business Media New York 2013

$[X^{n-}]_{1/n} \cdot mH_2O)$ are composed of sheets of edge-sharing polyhedra occupied by Ca^{II} and Al^{III} hydroxides, stacked with H_2O and charge-compensating anions in between [6,7]. These phases possess a positive structural charge and, therefore, have a strong potential for anion retardation in the interlayer. AFm phases occur in cementitious systems in various anionic forms: AFm-OH, AFm-CO_3, AFm-SO_4 (monosulfate) and AFm-Cl_2 (Friedel's salt). Thus, replacement of SO_4^{2-}, CO_3^{2-} and Cl^- in AFm by I^- is an uptake mechanism potentially capable of reducing the mobility of ^{129}I in cementitious materials.

AFm-I_2 was successfully synthesized in earlier studies by a co-precipitation method [8] and more recently by hydrothermal synthesis [4]. To the best of our knowledge, only one study has been reported on AFm phases containing a mixture of iodide and other anions [9]. In the latter work, the author showed that AFm-I_2 was forming a continuous solid solution series with AFm-SO_4 at 100 °C and also with AFm-Br_2 at 24 °C and at 150 °C, whereas two separate phases were observed for mixtures with chloride at 24 °C or at 220 °C.

The present study aims at elucidating the iodide uptake mechanisms by AFm-SO_4 under ambient conditions (23 ± 2 °C). A molecular-level understanding is aimed at using X-ray diffraction (XRD) and extended X-ray absorption fine structure (EXAFS) spectroscopy. Indeed, partial uptake of iodide by AFm may influence the ordering of anions in the interlayer galleries [10], with possible impact on phase stability and solubility. In addition, the solubility properties of pure AFm-I_2 and of its solid solution with AFm-SO_4 were assessed in order to be implemented in the thermodynamic database. Geochemical models involving radionuclide species are required for performance assessment of the deep-geological disposal of radioactive waste. To this end, the lack of solubility data and the poor understanding of the iodide uptake mechanism by AFm prevented any reliable predictions of the fate of ^{129}I in cementitious systems. The results of this study will therefore be useful for predicting the fate of iodide in the performance assessment of a repository for low- and intermediate-level radioactive waste.

2 Materials and Methods

2.1 Precipitation and Dissolution Experiments

A pure tricalcium aluminate C_3A ($Ca_3Al_2O_6$) was prepared in beforehand following the procedure given in Matschei et al. [11]. AFm-$(I_2)_x(SO_4)_{1-x}$ series were synthesized in a glove box under a N_2 atmosphere (O_2, CO_2 < 2 ppm) at 23 ± 2 °C by reacting C_3A with $CaI_2/CaSO_4$ at various ratios ($x = 0.0, 0.1, \ldots, 1.0$) in degassed Milli-Q® water at a liquid-to-solid ratio of 10. Replicates were prepared for two different equilibration times: 3 and 6 months. Dissolution experiments were started like co-precipitation experiments, but just with half of the initial water volume. After 3 months, the second half of water volume was added, after adding the amount of KOH required to obtain the same OH^- concentration as the one measured for each sample at the end of the 3 months equilibration time from the co-precipitation experiments. The dissolution samples were then equilibrated for 3 further months.

The solid phases were collected either by centrifugation or by filtration through 0.2-μm nylon filters. The solid phases were quickly washed with Milli-Q® water and then thoroughly rinsed with degassed absolute ethanol in order to remove all impurities of unreacted salts. The solid phases were dried over a saturated solution of $CaCl_2$ and stored in the glove box before solid characterization. The supernatant was kept for liquid phase analysis.

2.2 *Solid Phase Analysis*

Mineralogical determination of the synthesized products was made by XRD using an X'Pert Pro Phillips diffractometer. A diffracted beam monochromator with a Cu-Kα source and steps of 0.02°/s with a counting time of 2 s were used in a Bragg-Brentano mode. Solid stoichiometries were determined by chemical analysis. A total of 100 mg of solid phase was digested in suprapur® HNO_3 and analysed for Ca and Al using inductively coupled plasma-optical emission spectroscopy (ICP-OES). Replacement of anionic species in the interlayer with CO_3^{2-} allowed the SO_4^{2-} and I^- contents in the solid phases to be determined. To this aim, 100 mg of solid phase was placed in a 0.1 M K_2CO_3 solution. SO_4^{2-} and I^- were completely exchanged due to the large excess of CO_3^{2-} (~10 times) and the high stability of the AFm-CO_3 phase [11]. SO_4^{2-} and I^- released in the aqueous solution were then analysed by high-performance ion-exchange chromatography.

2.3 *Liquid Phase Analysis*

The OH^- concentration was measured using a pH electrode calibrated with fresh KOH solution (from 10^{-3} to 10^{-1} M). Dissolved Ca and Al concentrations were determined by ICP-OES. A second aliquot was analysed for I^- and SO_4^{2-} concentrations using high-performance ion-exchange chromatography. Charge balance errors of all samples lay within 1–15 %.

2.4 *EXAFS Spectroscopy*

Iodine K-edge bulk EXAFS spectra were collected at the Dutch-Belgian beamline (DUBBLE, BM26A) of the European Synchrotron Radiation Facility (ESRF) in Grenoble (France). The measurements were carried out in transmission mode using a liquid He cryostat system at a temperature of 15 K. The Si (111) monochromator angle was calibrated by assigning the energy of 33.169 keV to the first inflection point of the K-edge of a KI reference compound. All samples were measured in transmission mode with ionization chambers filled with 100 % Ar (I_0) and a 50–50 % He–Kr mixture (I_t). EXAFS data reduction was performed using the IFEFFIT software package (Athena) following standard procedures [12]. Spectra were normalized by fitting a third-degree polynomial to the EXAFS region. The energy was converted to photoelectron wave vector units ($Å^{-1}$) by assigning the origin E_0 to the first inflection point of the absorption edge. Radial structure functions (RSF) were obtained by Fourier transformation (FT) of the k^3-weighted $\chi(k)$ function between 2.9 and 12.0 $Å^{-1}$ using a Kaiser–Bessel window and an apodization parameter of 1.0.

2.5 Thermodynamic Modelling

The activities and chemical potentials of the solutes were determined using a Gibbs free energy minimization geochemical code GEM-Selektor (http://gems.web.psi.ch), with thermodynamic data given elsewhere [13–15]. Activity coefficients of aqueous species were computed with the built-in extended Debye–Hückel equation. All solubility products were calculated based on the measured ion concentrations.

3 Results and Discussion

3.1 Characterization of the Solid Solution Series

XRD revealed that all AFm-$(I_2)_x(SO_4)_{1-x}$ samples were crystalline. Small amounts of ettringite ($[Ca_6Al_3(OH)_{12}][SO_4^{2-}]_3 \cdot 26H_2O$) impurity were observed, especially at high SO_4^{2-} content. In order to estimate the I/SO_4 ratio in the solid phase, the ettringite contamination was estimated from the Ca/Al ratio measured in comparison to AFm and ettringite stoichiometries, i.e. Ca/Al = 2 and 3, respectively. Nonetheless, additional ^{125}I uptake experiments of trace-level iodide by ettringite were carried out in order to assess the possibility of iodide uptake. These experiments revealed that ettringite does not take up iodide and its presence should therefore not affect the AFm-I_2–AFm-SO_4 system. A single AFm phase was observed by XRD with the interlayer distance increasing continuously with I^-/SO_4^{2-}, indicating the presence of a solid solution series between AFm-I_2 and AFm-SO_4. Our results show that AFm-I_2 and AFm-SO_4 form a continuous solid solution series not only at 100 °C [9], but also at 23 ± 2 °C.

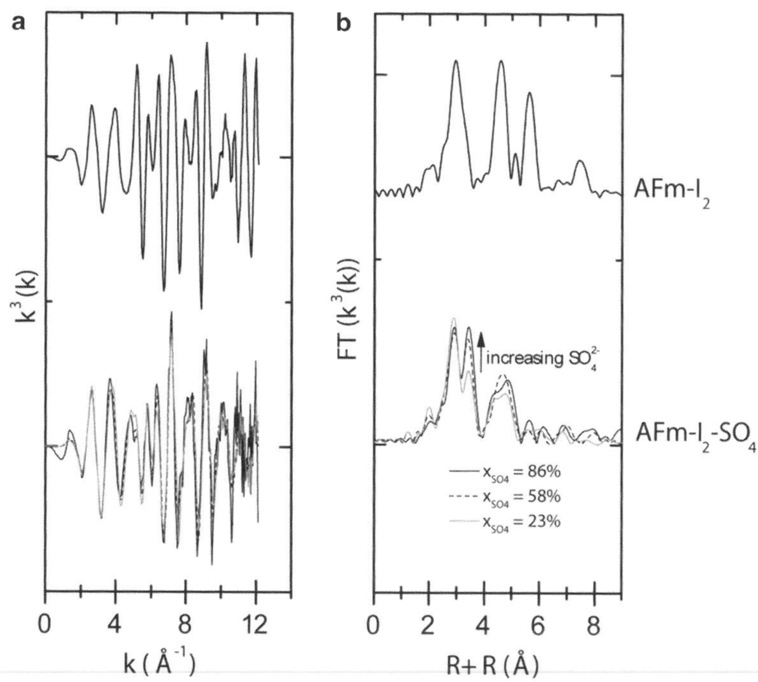

Fig. 6.1 Iodine K-edge spectra recorded at 15 K for pure AFm-I_2 and AFm (I_2, SO_4) samples with various $x_{(SO4)} = SO_4/(I_2 + SO_4)$ ratios in the solid: (**a**) k^3-weighted and (**b**) Fourier transform

3.2 *EXAFS*

The k^3-weighted spectra and their FT for AFm-I_2 and for AFm-$(I_2)_x(SO_4)_{1-x}$ are shown in Fig. 6.1. Heavy backscattering atoms like iodine seem to be present in the pure AFm-I_2 as indicated by high backscattering amplitudes at high k (Fig. 6.1a). In contrast, no enhanced amplitude was observed at high k in the case of AFm-$(I_2)_x(SO_4)_{1-x}$ (Fig. 6.1a). This indicates the absence of heavy backscattering atoms or the presence of a strong disorder, respectively. The Fourier transformed EXAFS spectra of AFm-$(I_2)_x(SO_4)_{1-x}$ showed a splitting of the first coordination shell with an additional peak at R + ΔR ~ 3.4 Å (Fig. 6.1b). Furthermore, the intensity of this shell at R + ΔR ~ 3.4 Å is enhanced with increasing SO_4^{2-} content (Fig. 6.1b).

The EXAFS data reveal that the coordination environment of iodine is significantly influenced by the presence of SO_4^{2-}. Both species may not only occupy the same interlayer, but they may even occupy neighbouring positions. This further implies that a complete mixing of I^- and SO_4^{2-} may occur within the interlayer of the AFm phases even at short range distances. Multi-shell fitting of the EXAFS spectra is planned to be performed in order to confirm the above findings. Theoretical scattering paths for the fit will be calculated with FEFF8.40 using structural models of pure AFm-I_2 and AFm-SO_4.

3.3 Solid Solution Solubility Products

Solubility data of the binary solid-solution system of AFm-$(I_2)_x(SO_4)_{1-x}$ are presented in terms of Lippmann diagrams [16]. The Lippmann total solubility product $\Sigma\Pi$ is defined as the sum of the solubility products of each end-member (introduced here in concentration scale). However, the Lippmann definition is strictly valid for substituting species that are identically charged. An extension of the Lippmann definition is required, if the exchanged species carry different charges. In the present study, iodide is charged (−1) and sulphate is charged (−2). The definition of the end-members can then follow two different conventions: the Gapon convention, where the stoichiometric coefficients of the matrix part $[Ca_4Al_2(OH)_{12}]$ of the end-members are identical: $Ca_4Al_2(OH)_{12}I_2$ and $Ca_4Al_2(OH)_{12}SO_4$, or the Vanselow convention, where the coefficients of the exchanged species are identical for both end-members (one I^- for one SO_4^{2-}): $Ca_2Al(OH)_6I$ and $Ca_4Al_2(OH)_{12}SO_4$. Both conventions result in two different definitions of $\Sigma\Pi$ and in different definitions of the end-member mole fraction in the solid solution. It can be shown that the end-member mole fraction is identical to the site mole fraction.

1. Gapon convention:

$$\Sigma\Pi = \left[Ca^{2+}\right]^4 \left[Al(OH)_4{}^-\right]^2 [OH^-]^4 \left([I^-]^2 + \left[SO_4{}^{2-}\right]\right) \tag{6.1}$$

$$x_{2\times I,aq} = \frac{[I^-]}{[I^-] + \left[SO_4{}^{2-}\right]} \tag{6.2}$$

$$x_{2\times I,s} = \frac{n_{Ca_4Al_2(OH)_{12}I_2}}{n_{Ca_4Al_2(OH)_{12}I_2} + n_{Ca_4Al_2(OH)_{12}SO_4}} = \frac{n_{I_2}}{n_{I_2} + n_{SO_4}} = \frac{\frac{n_I}{2}}{\frac{n_I}{2} + n_{SO_4}} \tag{6.3}$$

With n_I and n_{SO4}, number of moles of I and SO_4, respectively, in the solid.

2. Vanselow convention:

$$\Sigma\Pi = \left[Ca^{2+}\right]^2 \left[Al(OH)_4{}^-\right] [OH^-]^2 \left([I^-] + \left[SO_4{}^{2-}\right] \left[Ca^{2+}\right]^2 \left[Al(OH)_4{}^-\right] [OH^-]^2\right) \tag{6.4}$$

$$x_{I,aq} = \frac{[I^-]}{[I^-] + \left[SO_4{}^{2-}\right]} \tag{6.5}$$

$$x_{I,s} = \frac{n_{Ca_2Al(OH)_6I}}{n_{Ca_2Al(OH)_6I} + n_{Ca_4Al_2(OH)_{12}SO_4}} = \frac{n_I}{n_I + n_{SO_4}} \tag{6.6}$$

With n_I and n_{SO4}, number of moles of I and SO_4, respectively, in the solid.

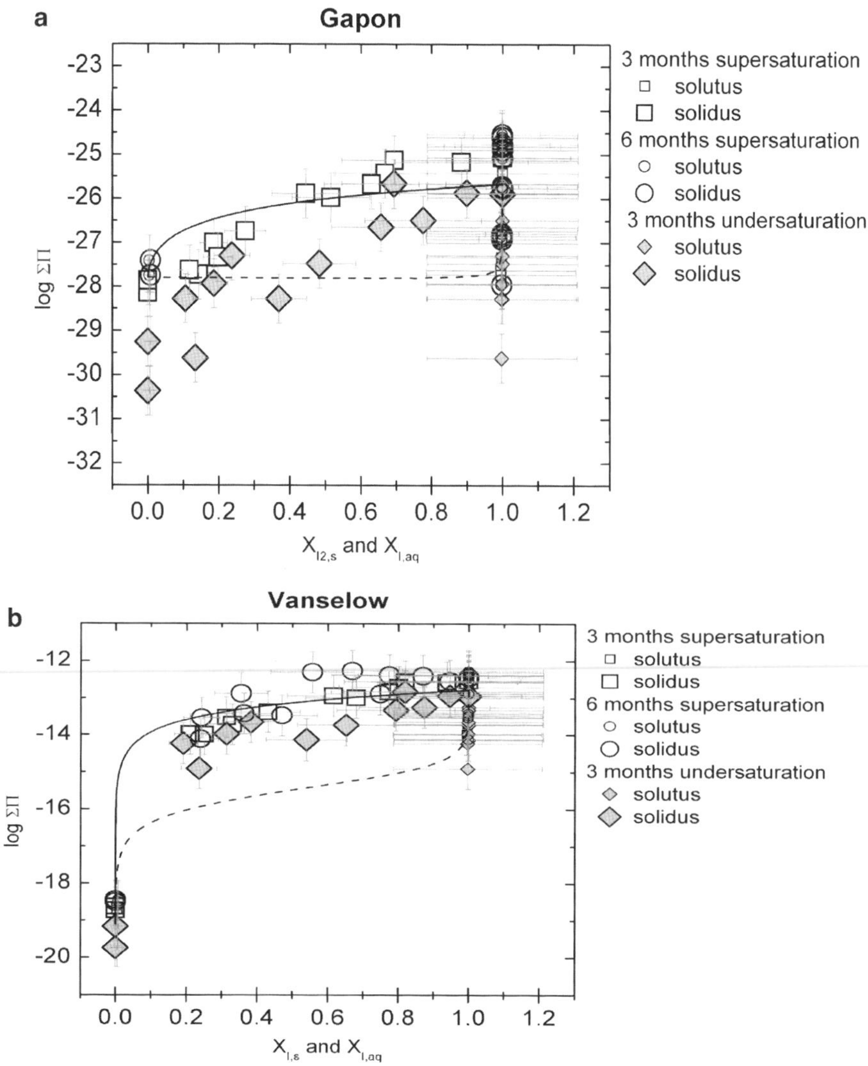

Fig. 6.2 Lippmann diagrams of the solid solution between AFm-I_2 and AFm-SO_4 (**a**) using Gapon convention and (**b**) using Vanselow convention. The *solid line* (solidus) and *dashed line* (solutus) represent the ideal solid solution model after assessing the solubility product of the end-members. The *symbols* are the experimental points after 3 or 6 months of co-precipitation (supersaturation) and after 3 months of dissolution (undersaturation) experiments

The solubility data were modelled using an ideal solid solution model. Theoretical values of $\Sigma\Pi$ were computed using Eqs. (6.1) or (6.4) from the output of GEM-Selektor speciation calculations, and of an ideal solid solution were calculated using the solubility products of the pure end-members and plotted as solidus [vs. Eq. (6.3) or (6.6)] and solutus [vs. Eq. (6.2) or (6.5)] curves against the experimental data in the case of the Gapon convention (Fig. 6.2a) and the Vanselow convention (Fig. 6.2b).

The results show that experimental solutus data are strongly displaced towards the iodine side ($x_{I,aq} \sim 1$), in a reasonable agreement with both models. The experimental solidus data indicate that equilibrium is approached from both supersaturation and undersaturation sides, matching the Vanselow convention closer. However, since the total solubility products from undersaturation experiments are systematically smaller than corresponding values from supersaturation experiments, one can conclude that full equilibration between aqueous and solid solution is not completely reached after 3–6 months ageing time. The equilibrium data are expected to lie in between the data determined in the undersaturation and supersaturation experiments. In conclusion, the experimental data appear to be consistent with a simple thermodynamic ideal solid solution model. Such models will eventually be used for predicting the ^{129}I concentrations in cementitious repository systems.

4 Conclusion

Formation of the continuous solid solution between AFm-I_2 and AFm-SO_4 was observed by XRD and supported by EXAFS analyses. A preliminary thermodynamic model indicates the formation of an ideal solid solution following the Vanselow convention. Our findings underpin the potential of AFm-SO_4 to act as a sink for ^{129}I.

Acknowledgements The authors thank the beamline scientists of the DUBBLE beamline (ESRF, Grenoble, France) for their technical support. Thanks are extended to Dr. Rainer Dähn and Dr. Marika Vespa for their support during EXAFS measurements and data analysis. Dr. Christine Taviot-Guého is greatly acknowledged for fruitful discussion and assistance during XRD data analysis. Dr. Urs Eggenberger and Christoph Wanner (UniBern) are acknowledged for assistance during the XRD measurements. Financial support was provided by the Helmholtz Virtual Institute for Advanced Solid-Aqueous Radio-Geochemistry (Germany) and by the National Cooperative for the Disposal of Radioactive Waste (Nagra), Switzerland.

References

1. Nagra (2002) Technical report. NTB 02-05. Nagra, Wettingen, Switzerland
2. Bonhoure I, Scheidegger AM, Wieland E, Dähn R (2002) Iodine species uptake by cement and CSH studied by I K-edge X-ray absorption spectroscopy. Radiochim Acta 90:647–651
3. Atkins M, Beckley AN, Glasser FP (1988) Influence of cement on the near-field environment and its specific interactions with uranium and iodine. Radiochim Acta 44–5:255–261
4. Rapin JP, Walcarius A, Lefèvre G, Francois M (1999) A double-layered hydroxide, $3CaO \times Al_2O_3 \times CaI_2 \times 10H_2O$. Acta Crystallogr Sect C Cryst Struct Commun 55:1957–1959
5. Walcarius A, Lefèvre G, Rapin JP, Renaudin G, Francois M (2001) Voltammetric detection of iodide after accumulation by Friedel's salt. Electroanalysis 13:313–320
6. Evans DG, Slade RCT (2006) Structural aspects of layered double hydroxides. In: Duan X, Evans DG (ed.), Layered Double Hydroxides. Springer-Verlag, Berlin Heidelberg 2005

7. Kuzel HJ, Pöllmann H (1991) Hydration of C_3A in the presence of $Ca(OH)_2$, $CaSO_4{\times}2H_2O$ and $CaCO_3$. Cement Concr Res 21:885–895
8. Brown DR, Grutzeck MW (1985) The synthesis and characterization of calcium aluminate monoiodide. Cement Concr Res 15:1068–1078
9. Kuzel H-J (1971) Zur Frage der Mischkristallbildung von Calciumaluminiumhydroxysalzen. Neues Jahrbuch für Mineralogie, Monatshefte 1971:477–491
10. Hou XQ, Bish DL, Wang SL, Johnston CT, Kirkpatrick RJ (2003) Hydration, expansion, structure, and dynamics of layered double hydroxides. Am Mineral 88:167–179
11. Matschei T, Lothenbach B, Glasser FP (2007) The AFm phase in Portland cement. Cement Concr Res 37:118–130
12. Ravel B, Newville M (2005) ATHENA, ARTEMIS, HEPHAESTUS: data analysis for X-ray absorption spectroscopy using IFEFFIT. J Synchrotron Radiat 12:537–541
13. Hummel W, Berner U, Curti E, Pearson FJ, Thoenen T (2002) Nagra/PSI thermodynamic database 01/01. Universal Publisher/uPublish.com, Parkland, FL
14. Lothenbach B, Matschei T, Moschner G, Glasser FP (2008) Thermodynamic modelling of the effect of temperature on the hydration and porosity of Portland cement. Cement Concr Res 38:1–18
15. Matschei T, Lothenbach B, Glasser FP (2007) Thermodynamic properties of Portland cement hydrates in the system $CaO\text{-}Al_2O_3\text{-}SiO_2\text{-}CaSO_4\text{-}CaCO_3\text{-}H_2O$. Cement Concr Res 37:1379–1410
16. Lippmann F (1980) Phase diagrams depicting the aqueous solubility of binary mineral systems. Neues Jahrbuch für Mineralogie (Abhandlungen) 139:1–25

Part III
Influence of External and Internal Factors on Long Term Properties of Cement-Waste Package and Cement Barriers

Chapter 7
Coupling Between Leaching and Mechanical Behaviour of Concrete

J.M. Torrenti, T. de Larrard, and F. Benboudjema

1 Introduction

Deep nuclear waste disposal facilities need to be studied over periods of one or more orders of magnitude greater than those of classical civil engineering. This leads us to consider some degradation phenomena that are most of the time not taken into account because of their slow kinetics. This is the case with calcium leaching for concrete structures exposed to water from the host rock in a deep underground storage. The calcium ion concentration in the interstitial pore solution of concrete decreases and this leads to a dissolution of the calcium bound in the skeleton of portlandite crystals $Ca(OH)_2$ and calcium-silica-hydrates (C-S-H). Calcium leaching leads to important changes in the concrete microstructure, and especially a large increase of porosity [1, 2]. In this chapter, the influence of leaching on the mechanical behaviour and then the influence of cracks on leaching are firstly considered. But the main coupled phenomenon that is to be considered here is tertiary creep. After a presentation of this phenomenon and its coupling with leaching, an example of a probabilistic approach to assess the lifetime of a simple structure is proposed.

J.M. Torrenti (✉)
Département Matériaux, Université Paris Est, IFSTTAR, Paris, France
e-mail: jean-michel.torrenti@ifsttar.fr

T. de Larrard • F. Benboudjema
LMT/ENS Cachan/CNRS UMR8535/UPMC/PRES UniverSud Paris, Cachan, France

F. Bart et al. (eds.), *Cement-Based Materials for Nuclear Waste Storage*,
DOI 10.1007/978-1-4614-3445-0_7,

2 Influence of Leaching on the Mechanical Behaviour of Concrete

2.1 At a Microscopic Scale

At a microscopic scale, microhardness measurements could be used to estimate the mechanical effect of leaching [3]. Figure 7.1 shows microhardness measurements on a leached cement paste (Fig. 7.1a) and on sound paste (Fig. 7.1b) with the same magnification. Vickers hardness could be deduced from these experiments by means of the imprint diagonals measurement. The measurements are normalised by means of the hardness of the sound paste. Figure 7.2 shows the distribution of

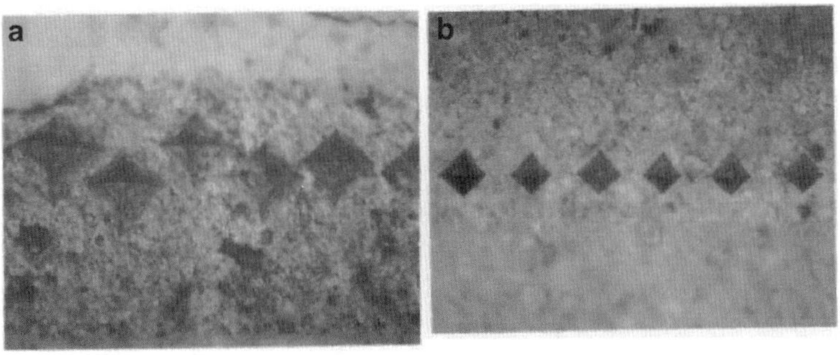

Fig. 7.1 Hardness imprints in (**a**) the leached zone (**b**) the sound zone of a cement paste

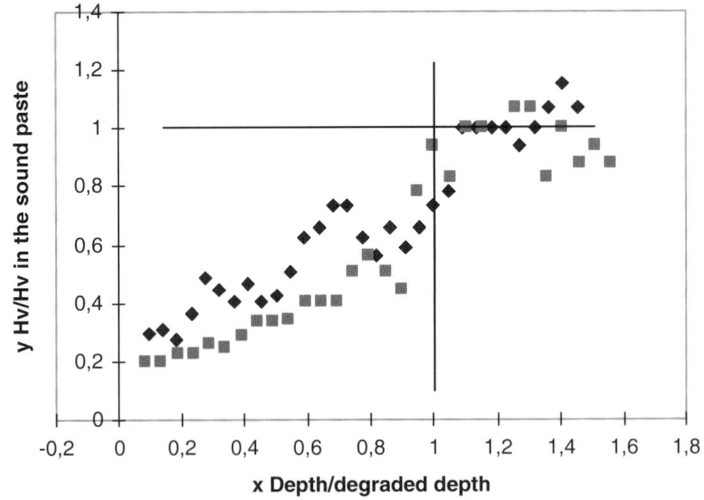

Fig. 7.2 Distribution of microhardness H_v (Vickers) in thickness of two degraded cement paste [3]. $x = 0$ corresponds to the surface of the sample that is leached

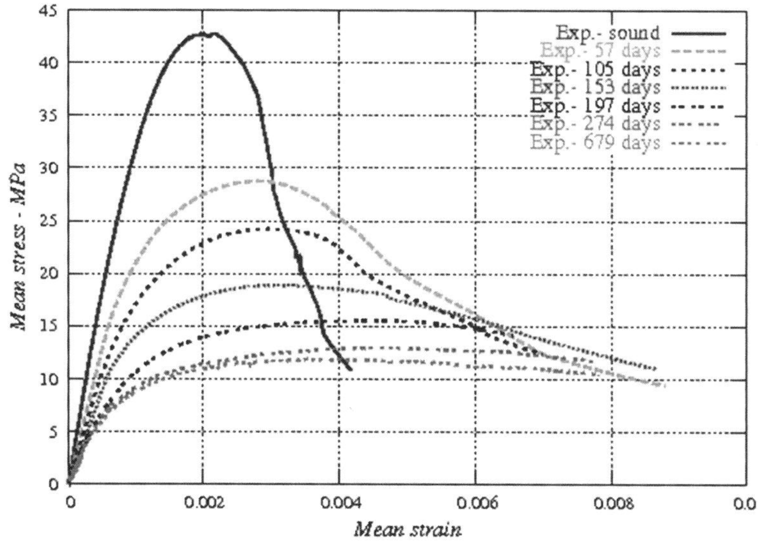

Fig. 7.3 Mean stress vs. mean strain curves at different degradation times for a concrete in compression [7]

microhardness H_v in the thickness of the degraded cement paste. These experiments could be used to relate the local mechanical properties such as the Young's modulus, the chemical softening and the plastic strain hardening to leaching [3].

2.2 *At a Macroscopic Scale*

After leaching, at a macroscopic scale, the mechanical properties of the leached material are also strongly affected by the evolution of the microstructure [4–7]. As leaching grows, a loss of strength and stiffness is observed, and a smoother post-peak behaviour with increasing irreversible straining is to be noted (Fig. 7.3). Cyclic loading suggests a plastic-like constitutive behaviour for the completely leached concrete in compression (Fig. 7.4). The behaviour of concrete is strongly influenced by leaching and evolves from damage (with small permanent strains) for sound concrete to plasticity (with large permanent strains, but also a decrease of mechanical properties as Young's modulus and strengths) for totally leached concrete. But in a real disposal, the loading will be sustained and not applied after leaching. That is why the coupling between creep and leaching will be discussed later in this chapter.

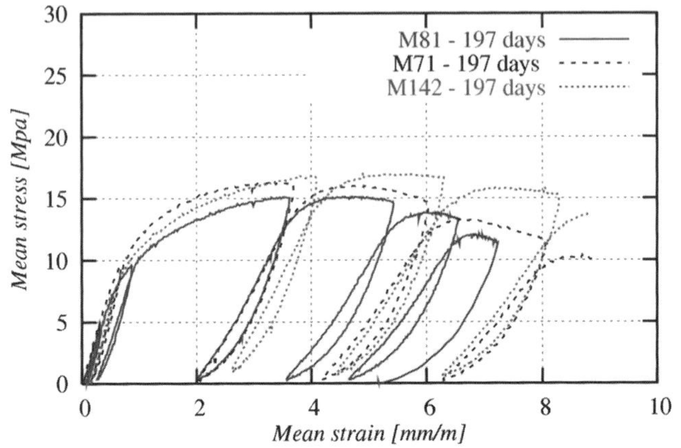

Fig. 7.4 Mean stress vs. mean strain curves under cyclic loading after 197 days of degradation [7]

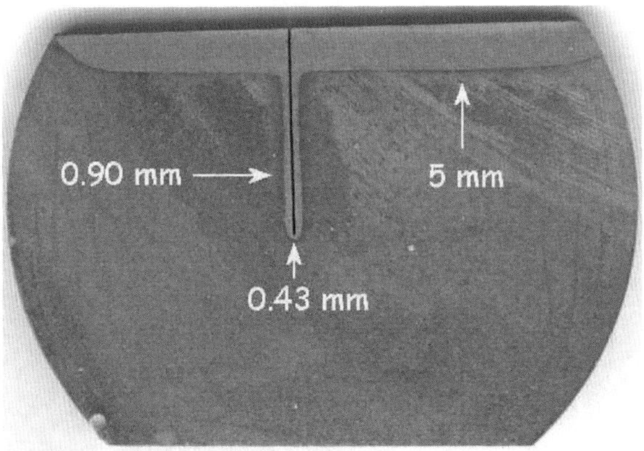

Fig. 7.5 Leaching of a crack (opening ≈ 350 μm) in a cement paste sample, by courtesy of CEA [8]

3 Influence of Cracks

It is very difficult to obtain an uncracked concrete structure. At a meso-scale, the restraint of cement paste shrinkage by the aggregates could be a source of microcracks. At a macro-scale, boundary conditions, that could also restrain delayed deformations, gradient of shrinkage (drying, carbonation, etc.) and the applied loading are possible sources of cracks in normal service conditions. That is why the influence of cracks on leaching must be considered.

An experiment performed at CEA by Peycelon [8] on a model crack with an opening of about 350 μm shows that, if there is no advection, the leaching at the crack tip is lower than at the surface of the sample (Fig. 7.5). Using similarity properties, Mainguy et al. [9] have shown that, in case of a pure diffusive process,

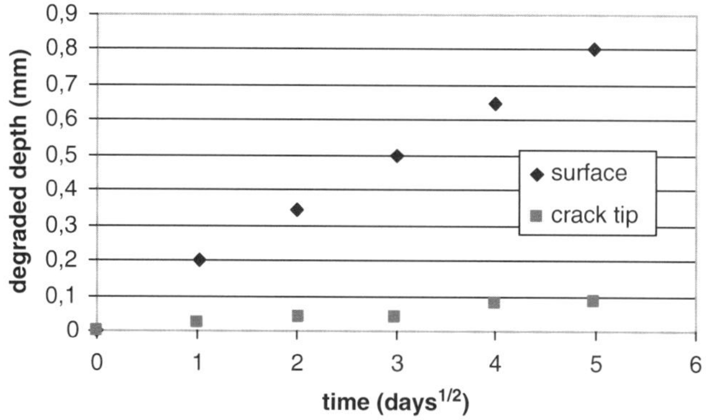

Fig. 7.6 Degraded depth at the surface and at the crack tip vs. the square root of time, experiment presented in Fig. 7.5 [8]

the kinetics of the degradation at the crack tip becomes a $t^{1/4}$ function after a period depending on the crack opening. In this case, the rate of degradation in the crack is lower than at the surface (Fig. 7.6). This was confirmed by Rougelot [10] using X-ray microtomography on a pre-cracked sample (using rapid drying) submitted to leaching: cracks generate only small local effects and do not affect the mean leaching depth. Note that Rougelot et al. [11] have also shown with the same technique that leaching induces microcracks in the leached zone because of both the mechanical properties decrease and the stresses generated by the decalcification shrinkage [12] restrained by aggregates. These microcracks do not affect significantly the leaching process.

4 Tertiary Creep

Finally the main coupling between leaching and the mechanical behaviour of concrete appears in the case of a sustained load. Indeed, because of leaching, the stress in the sound concrete increases and tertiary creep occurs. After a presentation of this phenomenon and its possible modelling, the case of a very simple structure will be presented taking into account variable properties.

4.1 A Coupling Between Creep and Cracking of Concrete

4.1.1 Experimental Evidence

Since Rüsch experiments [13], one knows that when concrete is subjected to a high sustained compressive load (stress higher than 80 % of the compressive strength), non-linear creep is observed and failure occurs after a while. This phenomenon is called tertiary creep. Several experimental results confirm this behaviour [14–16]. In tension some experimental evidence also exist using a unnotched sample loaded in tension [17, 18] or a notched bending beam [19] or a wedge splitting test [20]. And, at early age, tertiary creep could also be involved when shrinkage is restrained [21].

There is clearly a coupling between the high sustained load and the evolution of microcracking. Smadi and Slate [22] have shown that the length of microcracks evolves rapidly when the stress level is higher than 80 % of the compressive strength. Using acoustic emission, it is also possible to highlight microcracking during creep [23] or relaxation [20].

4.1.2 Modelling

Many modelling approaches have been proposed for tertiary creep of concrete. Bazant and Xiang [24], assuming that tertiary creep is only due to the cracks growing with time, uses a fracture mechanics approach coupled with creep to model this behaviour. Coupling between creep and cracking can be modelled by combining a viscoelastic and a viscoplastic model [25]. This model is built on Duvault–Lion approach in which a generalised Maxwell model is integrated. It reproduces with a satisfactory accordance linear viscoelastic creep, creep failure and rate effect on strength. Several damage models were also used. Li [14] proposed that the temporal variable can be explicitly introduced in the mechanical damage evolution law (in terms of damage rate). In this case, this approach is dissociated with linear viscoelastic creep strain. Challamel et al. [26] have developed a softening viscodamage model that could be viewed as a generalisation of a time-independent damage model and describes phenomena like relaxation, creep and rate-dependent loading. Recently, Sellier al. [27] have proposed a creep model taking into account a material consolidation and a creep induced damage that affects the creep potential. Mazotti and Savoia [28] proposed to model non-linear creep strain by introducing a stress rate reduction factor as a function of the damage variable in the solidification model [29]. Moreover, an effective strain is then defined for creep damage, replacing the equivalent strain (as defined by Mazars [30]) in damage evaluation for instantaneous loading case. The effective strain is defined as the sum of instantaneous damaged elastic strain with a fraction of creep strain. Omar et al. [18, 31] also used a similar approach.

Among these approaches, our model is based on the one proposed by Mazotti and Savoia [28], since the use of damage allows a simple and natural coupling with creep and, for the specific application of waste disposal, leaching through the use of the effective stress concept.

Creep Model

The model used for creep strain evolution is based on the microprestress theory proposed by Bazant and Xiang [24]. This model turns out to be particularly accurate

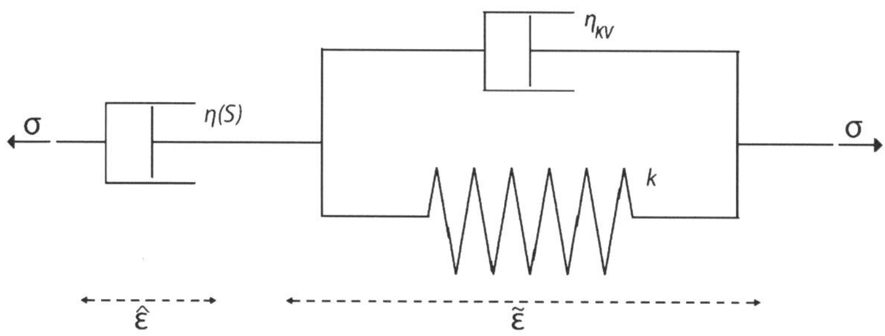

Fig. 7.7 Principle of the creep model

when considering a very long time creep and is in good accordance with experimental results (cf. the experimental campaign by Brooks [32] for 30 years creep). The microprestress solidification theory leads to introduce a dashpot whose viscosity depends on the microprestress so as to represent the microprestress relaxation mechanism. Therefore, this model is able to reproduce accurately ageing, even after hydration is completed. Microprestress is a stress level induced in the material by capillary pressure, evolving with time during the hydration process and with pore humidity. The strain evolution for the dashpot is given by Eq. (7.1), where $\tilde{\sigma}$ is the effective stress (assuming that only the uncracked part of the material undergoes creep, cf. section "Damage Model" for the definition of the effective stress), η is the viscosity of the dashpot and S is the microprestress.

$$\frac{\mathrm{d}\hat{\varepsilon}}{\mathrm{d}t} = \frac{\tilde{\sigma}}{\eta(S)} \tag{7.1}$$

The difficult part with this approach is to determine the viscosity evolution with microprestress as well as microprestress evolution with time. In the case of a saturated (meaning that there are neither drying nor wetting phenomena during the degradation and creep processes) and already aged structure (i.e. no hydration process is ongoing during the load application), microprestress and viscosity evolutions are quite simple [24] and leads to Eq. (7.2), where α is a creep parameter that remains constant and depends only on the material.

$$\frac{\mathrm{d}\hat{\varepsilon}}{\mathrm{d}t} = \tilde{\sigma}\frac{\alpha}{t} \tag{7.2}$$

The main disadvantage of the flow element is that it does not allow to represent the creep recovery: when the load is removed, no recovery can be modelled by the flow element [$\sigma = 0$ in Eq. (7.2)]. To represent recovery, a Kelvin–Voigt chain is added to the flow element in serial. Figure 7.7 shows, for a unidirectional problem, the complete creep model. The Kelvin–Voigt element is composed of a dashpot (viscosity η_{KV}) and an elastic spring (stiffness k) connected in parallel. Illston [33] has shown that the creep recovery lasts for about 1 week: thus, a value of 7 days for the characteristic time $\tau_{KV} = \eta_{KV}/k$ of the Kelvin–Voigt chain is used. The strain evolution for the Kelvin–Voigt element is given by Eq. (7.3), where $\tilde{\varepsilon}$ is the creep strain for the element under a stress $\tilde{\sigma}$. The expression of the total creep strain ε_c (sum of the 2 aforementioned strains) is given also in Eq. (7.3).

$$\frac{\tilde{\sigma}}{k} = \tilde{\varepsilon} + \tau_{KV}\frac{d\tilde{\varepsilon}}{dt} \quad \text{and} \quad \varepsilon_c = \tilde{\varepsilon} + \hat{\varepsilon} \tag{7.3}$$

Damage Model

The damage model proposed by Mazars [30] is used. In this model, a scalar mechanical damage variable D_c in compression is associated to the mechanical degradation process of concrete induced by the development of microcracks. It is defined as the ratio between the microcracks area to the whole material area. An effective stress $\tilde{\sigma}$ is then defined from the apparent stress σ applied on the whole material section. The relationship between apparent stress σ, effective stress $\tilde{\sigma}$, damage D_c, elastic stiffness tensor E, elastic strain ε_e, creep strain ε_c and total strain ε reads:

$$\sigma = (1 - D_c)\tilde{\sigma} \quad \text{and} \quad \frac{d\tilde{\sigma}}{dt} = E\frac{d\varepsilon_e}{dt} = E\left(\frac{d\varepsilon}{dt} - \frac{d\varepsilon_c}{dt}\right) \tag{7.4}$$

The Poisson ratio is assumed to be constant. The mechanical damage evolution is deduced from the Mazars equivalent tensile strain ε_{eq} defined in Eq. (7.5), where $\langle\varepsilon\rangle_+$ is the positive part of the strain tensor. The evolution law for the mechanical damage variable is given in Eq. (7.6), where ε_D is the damage tensile strain threshold, A and B are constant material parameters which control the hardening/softening branch in the stress–strain curve.

$$\varepsilon_{eq} = \sqrt{\langle\varepsilon\rangle_+ : \langle\varepsilon\rangle_+} \tag{7.5}$$

$$D_c = 1 - \frac{\varepsilon_D(1 - A)}{\varepsilon_{eq}} - \frac{A}{\exp[B(\varepsilon_{eq} - \varepsilon_D)]} \quad \text{if } \varepsilon_{eq} \geq \varepsilon_D \tag{7.6}$$

Coupling Between Damage and Creep

The coupling between creep and damage is taken into account twice. On the one hand, creep strains are considered to be driven by effective stresses (only the uncracked material creeps as shown in [34, 35], see section "Creep Model"). It slightly increases creep strains after damage occurs in compression, but it is not sufficient for correctly retrieving non-linear creep. On the other hand, the calculation of the equivalent strain [Eq. (7.5)] is modified as proposed by Mazotti and Savoia [28] and includes a part of creep strains through a material parameter β,

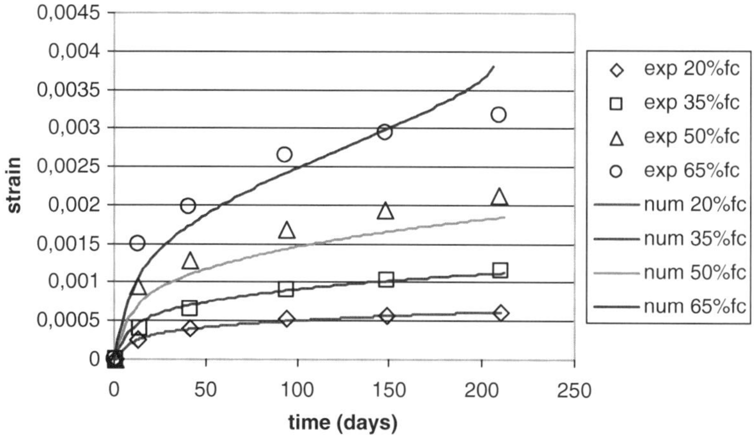

Fig. 7.8 Delayed strains for different stress to strength ratio (with coupling $\beta = 0.31$, Eq. (7.7)—experimental results from Roll [15])

which can be identified from non-linear basic creep data. Therefore, damage is also affected by basic creep strain. This allows to retrieve accurately non-linear basic creep evolutions for high stress levels [Eq. (7.7)]. Note that with this coupling, due to the choice of a non-asymptotic damage, failure will finally occur due to tertiary creep (but in a very long time—several thousands years—if the loading is lower than 40 % of the compressive strength).

$$\varepsilon_{\mathrm{eq}} = \sqrt{\langle \varepsilon + \beta \varepsilon_{\mathrm{c}} \rangle_+ : \langle \varepsilon + \beta \varepsilon_{\mathrm{c}} \rangle_+} \tag{7.7}$$

This model is compared to experimental results obtained by Roll [15]. The specimens have been loaded at an age of 28 days when the compressive strength was equal to 42 MPa. The levels of loading were equal, respectively, to 20, 35, 50 and 65 % of the compressive strength. The lower level is used to calibrate the parameters of the creep model (for this level of loading, there is no cracking, only primary creep occurs). Introducing a coupling with $\beta = 0.31$, the coupling between damage and creep allows to fit the experimental results (Fig. 7.8).

4.2 *Tertiary Creep and Leaching*

4.2.1 Modelling of Leaching

Considering that the dissolution is instantaneous (local equilibrium) and that only calcium species are to be taken into account, the leaching of a cement paste can be described by the mass balance equation of calcium [Eq. (7.8)], as it was proposed by Buil et al. [36], where S_{Ca} is the solid calcium concentration, C_{Ca} is the liquid calcium concentration, D is the calcium effective diffusivity in porous material and ϕ is the porosity. One can recognise the two main phenomena involved in the leaching process: on the one hand, the solid and liquid phases of calcium in the cement paste and in the porous solution are in chemical equilibrium; on the other hand, the ionic calcium species diffuse through the material porosity. The non-linearity of this equation is mainly due to the diffusivity which depends on the porosity, itself depending on the solid calcium concentration.

$$\frac{\partial(C_{\mathrm{Ca}}\phi)}{\partial t} = -\mathrm{div}[-D(\phi)\mathrm{grad}(C_{\mathrm{Ca}})] - \frac{\partial S_{\mathrm{Ca}}}{\partial t} \tag{7.8}$$

In a semi-infinite medium, because leaching is driven by diffusion, the degraded depth x_{d}, corresponding to the dissolution front of portlandite, is a function of the square root of time:

$$x_{\mathrm{d}} = k\sqrt{t} \tag{7.9}$$

4.2.2 Coupled Mechanical and Chemical Damage

The coupling between mechanical behaviour and chemical degradation of concrete has been studied in details (e.g. [5, 7, 37–39]). Gérard et al. [40] proposed a simplified modelling for chemical alteration and a coupling with the mechanical damage proposed by Mazars [30]. The coupling appeared through the evolution of the coefficient of diffusion depending on the damage variable. The model proposed by Ulm et al. [3] is based on the theory of porous media and couples the hydrates dissolution with the evolution of porosity and strain. The model of Kuhl et al. [41, 42] proposes to consider three components of porosity (initial porosity, porosity due to chemical degradation and porosity due to mechanical damage). Recently, Sellier et al. [43] have proposed a non-linear viscoelastic model in which the non-linearities are due to damage or creep. In this model, the mechanical properties are functions of the calcium concentration in the solid phase.

Studies led on degraded materials [7] have quantified the loss of residual mechanical properties for leached concrete. This decrease mainly depends on the mineral admixtures in the concrete mix design and on the degraded area which is considered (dissolution of portlandite only does not lead to very significant decrease of the compressive strength, whereas dissolution of C-S-H does). Here, for sake of simplicity, it will be considered that the leached area does not sustain any load. Indeed, considering more important residual mechanical properties for the degraded areas would lead to introduce more input parameters in our model, and, therefore, more variability. Such a strong hypothesis on the mechanical behaviour of the degraded area has already been used by Torrenti et al. [39] and leads to an under-estimation of the lifetime of the considered structure (which is in acceptable accordance with a view of preserving the safety of concrete structures).

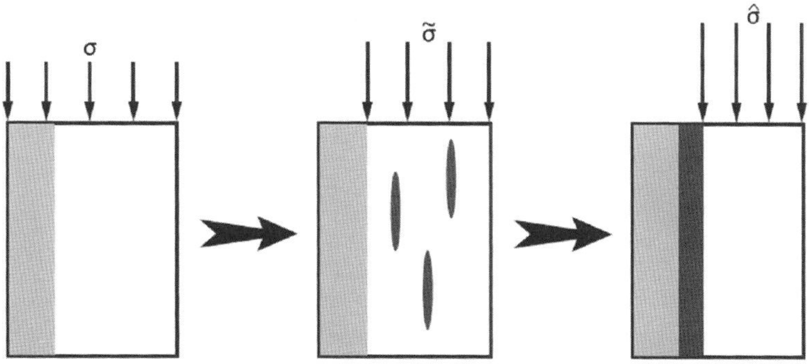

Fig. 7.9 Definition of effective stress on damaged material: no stress is sustained by the chemically damaged section (*light grey*), so the real stress σ becomes $\tilde{\sigma}$ on the sound material (*white*). Due to mechanical damage, cracks in the sound material (*deep grey*) reduce the material section to sustain the load, and thus introduce an effective stress $\hat{\sigma}$ on the uncracked sound material

4.2.3 Chemical Damage

A chemical damage variable D_χ is associated to the leaching degradation process of concrete and is defined as the ratio of the degraded area over the whole material section. Figure 7.9 shows how the effective stress $\tilde{\sigma}$ is defined from the apparent stress σ applied on the whole material section (included the degraded area): the stress that is actually applied on the chemically sound area is increased according to the same ratio that the section is reduced. When a mechanical loading exists, the effective stress $\hat{\sigma}$ applied on the uncracked and chemically sound material is thus deduced from the apparent stress σ and the chemical and mechanical damage variables D_χ andD_c [Eq. (7.10)]:

$$\hat{\sigma} = \frac{\sigma}{(1 - D_\chi)(1 - D_c)} \tag{7.10}$$

4.2.4 Example of Application: Creep Coupled to Leaching [39]

When a loaded cylindrical sample (radius R) is also submitted to leaching (here accelerated leaching using an ammonium nitrate solution), the stress σ in the sound zone increases during the test:

$$\sigma(t) = \sigma(t_0)\frac{\pi R^2}{\pi\left(R - k\sqrt{t}\right)^2} \tag{7.11}$$

and finally tertiary creep occurs (Fig. 7.10). Using the coupling between chemical and mechanical damage, the experimental behaviour could be modelled.

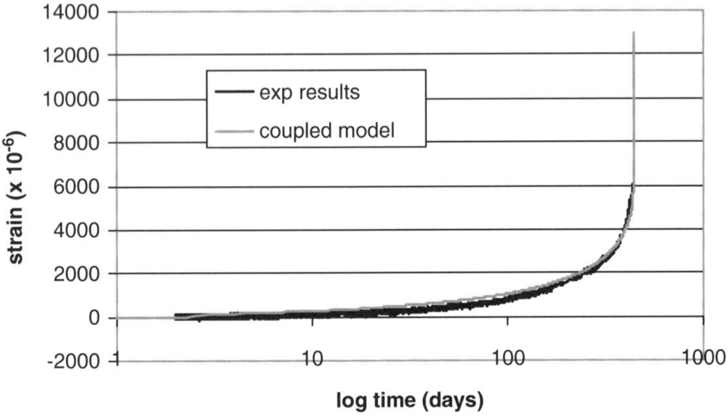

Fig. 7.10 Comparison between experimental strains in a leaching and creep test and the coupled chemical–mechanical model

Table 7.1 Expected value and standard deviation for the seven variable input parameters of the model

	Expected values	Standard deviation
ϕ [−]	0.129	0.010
τ [−]	0.134	0.020
E [ΓΠα]	45.0	4.5
υ [−]	0.178	0.044
β [−]	0.15	0.03
ε_{Δ} [∝μ/μ]	109	16
ς [−]	0.95	0.10

4.3 Probabilistic Approach [44]

We consider here a column made of concrete (reinforcement is not considered) in pure compression subjected to leaching due to pure water. The radius of the column is equal to 20 cm. With the values of the parameters which are given in Table 7.1, tertiary creep is obtained. The evolution of damage is analysed: in a first time, only chemical damage occurs; then the coupling between chemical and mechanical damage appears and finally failure occurs due to mechanical damage (Fig. 7.11).

Considering Beta(2,2) distributions for each of the seven parameters of the model, 1,000 corresponding sets of input parameters were randomly generated according to the Latin Hypercube Sampling principle [45], optimised thanks to a simulated annealing algorithm [46]. The expected value and standard deviation for each variable parameter of the model are presented in Table 7.1.

The expected value for the lifetime of the concrete structure, evaluated over the 1,000 realisations, is 980 years. The observed standard deviation is 393 years, which corresponds to a coefficient of variation of 40 %. Of course, this important value for the coefficient of variation must be compared to the variability of the nine input parameters, whose coefficient of variation ranges between 10 and 25 %.

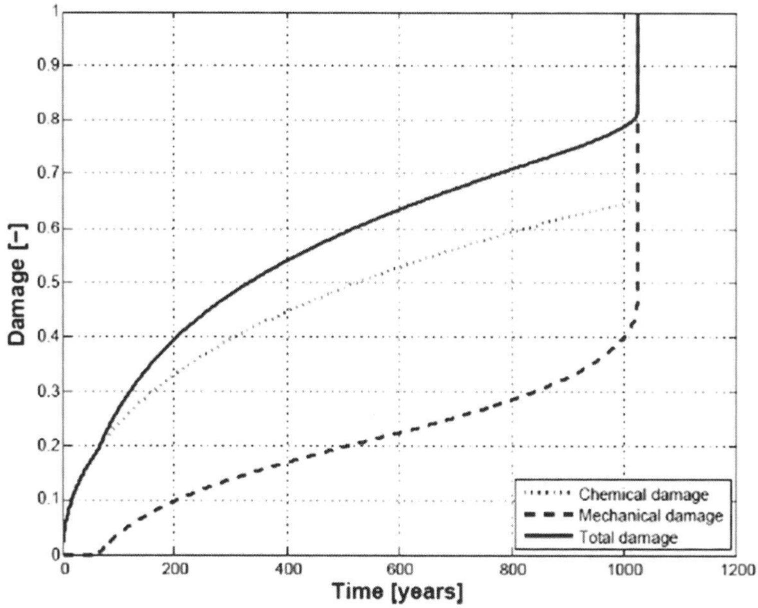

Fig. 7.11 Total, mechanical and chemical damage time evolutions obtained in pure water

Thus, the coupled problem we are dealing with leads to an increasing uncertainty. It is interesting to note that lifetime follows a log-normal distribution, whereas all input parameters are distributed according to a Beta law. Figure 7.12 shows the lifetime distribution that has been calculated on the 1,000 realisations, which can be compared to a log-normal distribution.

5 Conclusions

Leaching of concrete by water affects its microstructure and then its mechanical behaviour: the compressive strength and the stiffness decrease with leaching. And the behaviour of concrete evolves from damage (with small permanent strains) for sound concrete to plasticity (with large permanent strains) for totally leached concrete. In case of a cracked concrete, whatever is the source of the cracks (leaching itself or internal or external stresses), if there is no water advection, leaching at the crack tip is very low compared to leaching at the surface of concrete.

Finally, the main coupling is between leaching and the mechanical behaviour of concrete appears in the case of a sustained loading. Indeed, because of leaching, the stress in the sound concrete increases and tertiary creep occurs. This phenomenon is due to a coupling between creep and damage. It leads to failure in case of a high

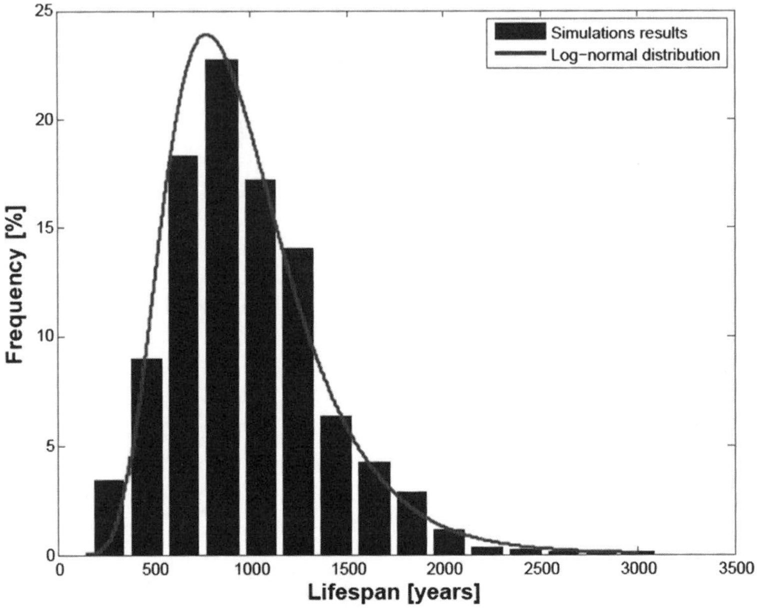

Fig. 7.12 Lifetime distribution for the 1,000 concrete structures simulated: accordance between the simulations results and a log-normal distribution

level of loading compared to the strength of concrete. This coupling could be taken into account by means of a creep model coupled to a mechanical damage model.

When concrete is also submitted to leaching, a chemical coupling has to be taken into account. A model for calcium leaching based on the calcium mass balance equation is used. Assuming that only the sound zone of concrete is able to sustain stresses, the damage evolution with time of a concrete column under pure compression and exposed to leaching and creep at the same time could be forecast. Assuming a Beta distribution of all the parameters of the model, a probabilistic study shows that the lifetime of this simple structure is distributed according to a log-normal law.

Acknowledgements The authors thank Patrick Le Bescop for his help concerning the reference on experiments at CEA.

References

1. Adenot F (1992) Durabilité du béton: caractérisation et modélisation des processus physiques et chimiques de dégradation du ciment, PhD thesis, Université d'Orléans (in French)
2. Gérard B (1996) Contribution des couplages mécaniques-chimie-transfert dans la tenue à long terme des ouvrages de stockage de déchets radioactifs, PhD thesis, ENS Cachan (in French)

3. Ulm FJ, Torrenti JM, Adenot F (1999) Chemoporoplasticity of calcium leaching in concrete. J Eng Mech 125(10):1200–1211
4. Bernard F, Kamali-Bernard S, Prince W (2008) 3D multi-scale modelling of mechanical behaviour of sound and leached mortar. Cement Concr Res 38:449–458
5. Carde C, Francois R, Torrenti JM (1996) Leaching of both calcium hydroxyde and CSH from cement paste: modeling the mechanical behavior. Cement and concrete research 26(8):1257–1268
6. Huang B, Qian C (2011) Experiment study of chemo-mechanical coupling behavior of leached concrete. Constr Build Mater 25:2649–2654
7. Nguyen VH, Colina H, Torrenti JM, Boulay C, Nedjar B (2007) Chemomechanical coupling behaviour of leached concrete. Part 1: experimental results. Nucl Eng Des 237:2083–2089
8. Richet C, Galle C, Le Bescop P, Peycelon H, Bejaoui S, Tovena I, Pointeau I, L'Hostis V, Lovera P (2004) Synthèse des connaissances sur le comportement à long terme des bétons – Application aux colis cimentés, rapport CEA-R-6050, 2004 (in French)
9. Mainguy M, Ulm FJ, Heukamp FH (2001) Similarity properties of demineralization and degradation of cracked porous materials. Int J Solid Struct 38:7079–7170
10. Rougelot T (2008) Etude expérimentale multi-échelles des couplages hydriques, mécaniques et chimiques dans les matériaux cimentaires, PhD thesis, Université de Lille (in French)
11. Rougelot T, Burlion N, Bernard D, Skoczylas F (2010) About microcracking due to leaching in cementitious composites: X-ray microtomography description and numerical approach. Cement and concrete research 40(2):271–283
12. Chen JJ, Thomas JJ, Jennings HM (2006) Decalcification shrinkage of cement paste. Cement Cement and concrete research 36:801–809
13. Rüsch H (1960) Researches toward a general flexural theory for structural concrete. ACI J 32(1):1–28
14. Li Z (1994) Effective creep Poisson's ratio for damages concrete. Int J Fract 66:189–196
15. Roll R (1964) Long time creep-recovery of highly stressed concrete cylinders, ACI SP-9, Symposium on creep, Portland Cement Association, Detroit, pp 115–128
16. Smadi MM, Slate FO, Nilsson AH (1987) Shrinkage and creep of high, medium and low strength concrete, including overloads. ACI Mater J 84(3):224–234
17. Reinhardt H-W, Rinder T (2006) Tensile creep of high-strength concrete. J Adv Concr Technol 4(2):277–283
18. Reviron N (2009) Etude du fluage des bétons en traction. Application aux enceintes de confinement des centrales nucléaires à eau sous pression, PhD thesis, ENS de Cachan (in French)
19. Carpinteri A, Valente S, Zhou FP, Ferrara G, Melchiorri G (1997) Tensile and flexural creep rupture tests on partially damaged concrete specimens. Mater Struct 30:269–276
20. Denarié E, Cécot C, Huet C (2006) Characterization of creep and crack growth interactions in the fracture behavior of concrete. Cement and concrete research 36:571–575
21. Briffaut M, Benboudjema F, Nahas G, Torrenti JM (2011) Numerical analysis of the thermal active restrained shrinkage ring test to study the early age behavior of massive concrete structures. Eng Struct 33(4):1390–1401. doi:10.1016/j.engstruct.2010.12.044
22. Smadi MM, Slate FO (1989) Microcracking of high and normal strength concretes under short and long term loadings. ACI Mater J 86(2):117–127
23. Rossi P, Godart N, Robert JL, Gervais JP, Bruhat D (1994) Investigation of the basic creep of concrete by acoustic emission. Mater Struct 27(9):510–514
24. Bazant ZP, Xiang Y (1997) Crack growth and life time of concrete under long time loading. J Eng Mech 123(4):350–358
25. Berthollet A, Georgin JF, Reynouard JM (2004) Fluage tertiaire du béton en traction. Revue européenne de Génie Civil 8(2–3):235–260
26. Challamel N, Lanos C, Casandjian C (2005) Creep damage modelling for quasi-brittle materials. Eur J Mech Solid 24:593–613

27. Sellier A, Multon S, Buffo-Lacarrière L (2011) Non linear creep modelling, LMDC report no. 03-201
28. Mazotti C, Savoia M (2003) Non linear creep damage model for concrete under uniaxial compression. J Eng Mech 129(9):1065–1074
29. Bazant ZP, Prasannan S (1989) Solidification theory for concrete creep. I. Formulation. J Eng Mech 115(8):1691–1703
30. Mazars J (1986) A description of micro and macroscale damage of concrete. Eng Fract Mech 25:729–737
31. Omar M, Pijaudier-Cabot G, Loukili A (2004) Etude comparative du couplage endommagement – fluage. Revue Française de Génie Civil 8:457–482
32. Brooks JJ (2005) 30-year creep and shrinkage of concrete. Mag Concr Res 57(9):545–556
33. Illston JM (1965) The components of strains in concrete under sustained compressive stress. Mag Concr Res 17(50):21–28
34. Benboudjema F, Meftah F, Torrenti JM (2005) Interaction between drying, shrinkage, creep and cracking phenomena in concrete. Eng Struct 27:239–250
35. Benboudjema F, Torrenti JM (2008) Early age behaviour of concrete nuclear containments. Nucl Eng Des 238(10):2495–2506
36. Buil M, Revertégat E, Oliver J (1992) A model of the attack of pure water or undersaturated lime solutions on cement, vol 2nd. American Society for Testing and Materials, Philadelphia, PA, pp 227–241
37. Bangert F, Kuhl D, Meschke G (2001) Finite element simulation of chemo-mechanical damage under cyclic loading conditions. In: de Borst R, Mazars J, Pijaudier-Cabot G, van Mier J (eds) Fracture mechanics of concrete structures, vol 1. Balkema, Rotterdam, pp 145–152
38. Saetta A, Scotta R, Vitaliani R (1998) Mechanical behavior of concrete under physical-chemical attacks. J Eng Mech (ASCE) 124:1100–1109
39. Torrenti JM, Nguyen VH, Colina H, Le Maou F, Benboudjema F, Deleruyelle F (2008) Coupling between leaching and creep of concrete. Cement Concr Res 38(6):816–821
40. Gérard B, Pijaudier-Cabot G, Laborderie C (1998) Coupled diffusion-damage modelling and the implications on failure due to strain localization. Int J Solid Struct 35(31–32):4107–4120
41. Kuhl D, Bangert F, Meschke G (2003) Coupled chemo-mechanical deterioration of cementitious materials - Part 1: modeling. Int J Solid Struct 41:15–40
42. Kuhl D, Bangert F, Meschke G (2003) Coupled chemo-mechanical deterioration of cementitious materials – Part 2: numerical method and simulation. Int J Solid Struct 41:41–67
43. Sellier A, Buffo-Lacarrière L, El Gonnouni M, Bourbon X (2010) Behavior of HPC nuclear waste disposal structures in leaching environment. Nucl Eng Des. doi:10.1016/j.nucengdes.2010.11.002
44. de Larrard T, Benboudjema F, Colliat JB, Torrenti JM, Deleruyelle F (2010) Uncertainty propagation on damage evolution of a concrete structure submitted to coupled leaching and creep. EJECE 14(6–7):891–921
45. McKay MD, Conover WJ, Beckman RJ (1979) A comparison of three methods for selecting values of input variables in the analysis of output from a computer code. Technometrics 21:239–245
46. Kirkpatric S, Gelatt CD, Vecchi MP (1983) Optimization by simulated annealing. Science 220 (4598):671–680

Chapter 8
Coupled Mechanical and Chemo-Transport Model for the Simulation of Cementitious Materials Subjected to External Sulfate Attack

B. Bary, N. Leterrier, E. Deville, and P. Le Bescop

1 Introduction

External sulfate attacks (ESA) in cementitious materials classically designate the degradation due to the penetration of sulfate ions within the connected porosity and their reaction with the hydrated products. In most situations these reactions appear when the material is in contact with water containing the aggressive species, which means that it is also subjected to leaching. In these conditions, the porosity is generally considered as saturated. It is commonly accepted that the main reactions in ESA, in addition to decalcification, involve the sulfate ions and monosulfoaluminate (AFm) initially present in the solid phase to precipitate into ettringite (8.1a), and possibly these sulfate ions and calcium ones to form gypsum (8.1b).

$$3CaOAl_2O_3CaSO_412H_2O + 2Ca^{2+} + 2SO_4^{2-} + 20H_2O \leftrightarrow 3CaOAl_2O_33CaSO_432H_2O \qquad (8.1a)$$

$$Ca^{2+} + SO_4^{2-} + 2H_2O \leftrightarrow CaSO_42H_2O \qquad (8.1b)$$

These phenomena may be accompanied by significant free expansions leading to severe mechanical damage. Although the exact origin of the pressure causing such macroscopic expansions is controversial, ettringite precipitation from AFm is generally recognized as the major cause of macroscopic swelling [1]. For some

---

B. Bary (✉) • N. Leterrier • E. Deville • P. Le Bescop
Commissariat à l'Energie Atomique et aux Energies Alternatives, DEN/DANS/DM2S/SFME, Gif-sur-Yvette, France
e-mail: benoit.bary@cea.fr

F. Bart et al. (eds.), *Cement-Based Materials for Nuclear Waste Storage*,
DOI 10.1007/978-1-4614-3445-0_8, © Springer Science+Business Media New York 2013

authors, the internal pressure is mainly due to the growing of small AFm crystals embedded in C-S-H matrix (e.g., [1]). In this case, the expansion process is assumed to be due to crystallization pressures resulting from the interaction between the solid product of chemical reactions (ettringite) and the surrounding matrix of cement paste [2]. This approach then suggests that chemical energy is converted into mechanical work whenever the solid product is confined, the driving forces being the activities of the reactants (supersaturation) in the pore solution. The crystallization pressure p_c is generally expressed by the Correns's equation:

$$p_c = \frac{RT}{\bar{V}_{\text{crys}}} \ln\left(\frac{Q_{\text{reac}}}{K_{\text{reac}}}\right), \quad \text{with } Q_{\text{reac}} = \prod_i a_i^{\upsilon_i^{(\text{reac})}} \tag{8.2}$$

where $\bar{V}_{\text{crys}}$ is the molar volume of the growing crystal, R is the gas constant, T is the temperature, a_i and $\upsilon_i^{(\text{reac})}$ the activity and the stoichiometric coefficient of species i in solution, respectively, and K_{reac} is the equilibrium constant of the chemical reaction. The main assumptions underlying (8.2) are that the growing crystal is supposed to be in hydrostatic state of stress and that the curvature radius of the crystal surface is large enough to neglect the contribution of the surface free energy (see, e.g., [3, 4]). Confinement and supersaturation of the solution are the two necessary conditions for this pressure to arise [1].

We study in this chapter the macroscopic free expansions of mortar samples submitted to sodium sulfate solutions. A coupled chemo-mechanical approach is proposed to simulate the behavior of these samples, with a particular attention paid to the mechanical modeling of the macroscopic expansions. The numerical platform ALLIANCES allowing to perform reactive transport calculations coupled with mechanics is used for the simulations [5]. Numerical results are compared to experimental data in terms of mineralogical profiles at given time and temporal evolutions of expansions.

2 Estimation of the Parameters Evolution

In this study, we estimate the evolution of diffusive and mechanical properties of hardened cement pastes (HCP) and mortars by making use of effective medium approximation schemes. A precise description of their microstructure is then required. The proposed two-scale description of HCP is based on the introduction of two microstructural levels [6–8]. The first one ranges from a few tens of micrometers to about 100 μm and is typical of the size of cement grains, while the second one ranges from 0.2 μm to a few tens of micrometers and corresponds to the scale of hydration products, such as portlandite (CH) or aluminate phases, and of capillary pores. The level I corresponding to the biggest scale of HCP represents the hydrated cement grains. During the hydration of cement particles, we assume that two layers defined as inner and outer layers form successively from cement grains surface. An anhydrous part of the cement particles remains after full hydration and constitutes a core surrounded by these two heterogeneous layers, all being spherical.

The generalized self-consistent scheme (GSCS) is applied to work out this microstructure [7]. The level II corresponds to the scale of the heterogeneities of the two layers. Two different types of C-S-H are introduced. In the inner layer, the CSH^{int} behaves as a matrix phase embedding inclusions of CH and aluminates. Similarly, the external layer has the same matrix-inclusion type morphology, where CSH^{ext} is the matrix phase. We assume that the capillary pores are distributed in the outer coatings. The properties of these two layers are estimated by means of the Mori-Tanaka (MT) method. The proposed description of the mortar microstructure involves the interfacial transition zone (ITZ) which is assumed to be a homogeneous phase. A doubly coated spheres model is again adopted to represent mortars: the core, the intermediary layer, and the external one are occupied by sand particles, ITZ and HCP, respectively. The macroscopic properties of mortars are also estimated by the GSCS.

3 Mechanical Behavior of the Cement Paste Subjected to ESA

In the approach developed in [9], the mechanical aspects are coupled with chemo-transfer ones through an analogy with poroelasticity. The crystallization pressure as defined in (8.2) and generated by confined growing crystals of AFm plays the role of the pore pressure. This model has been successfully applied to the determination of cracking initiation consecutive to ESA in cement pastes. However, it has been shown in [10] that the application of this approach leads to very small macroscopic free expansions, very far from the experimental measurements (about two orders of magnitude lower). The reason is that the calculated crystallization pressures are not sufficient to generate strains values comparable to experimental ones with a purely elastic approach. We then suspect that the observed expansions may also result at least partly from other phenomena such as microcracking and creep (see, e.g., [11]). This suggests that an alternative approach has to be developed for more accurately describing the free expansions. We propose to apply the model developed in [12] (also used in [11] and [13], and appearing in a slightly different way in [14]) and based on the introduction of a macroscopic bulk strain $\boldsymbol{\varepsilon}^*$ for reproducing the expansions due to ettringite formation. The hypothesis is made that ettringite first precipitates in larger pores without generating expansions. After a fraction of these pores quantified by $\beta\phi^0$, with ϕ^0 the initial porosity and β a parameter to identify, are filled by ettringite, macroscopic strains appear. They are expressed by:

$$\boldsymbol{\varepsilon}^* = 1/3\langle \phi_{\text{Ett}} - \phi^0_{\text{Ett}} - \phi_{\text{AFm}} - \beta\phi^0 \rangle_+ \boldsymbol{1} = \varepsilon^*_{\text{v}} \boldsymbol{1} \tag{8.3}$$

with ϕ^0_{Ett} the initial volume fraction of ettringite, $\boldsymbol{1}$ the second-order identity tensor, and $\langle\cdot\rangle_+$ the Macauley brackets. The macroscopic behavior law is then written in the form:

$$\boldsymbol{\sigma} = \boldsymbol{K} : (\boldsymbol{\varepsilon} - \boldsymbol{\varepsilon}^*) - \alpha_{\mathrm{AFt}} p_{\mathrm{c}} \boldsymbol{1} \tag{8.4}$$

where $\boldsymbol{K}$, σ, $\boldsymbol{\varepsilon}$, p_{c}, and α_{AFt} are the fourth-order rigidity tensor, the stress tensor, the strain tensor, the crystallization pressure as defined in (8.2), and the interaction coefficient, respectively. The expression of α_{AFt} is detailed in [10]. The model defined by (8.3) and (8.4) is implemented in the platform ALLIANCES. In this first stage, no damage or microcracking phenomena are considered.

4 Simulation Results

As mentioned in the introduction, the model is applied to the simulation of laboratory tests performed at EMPA [15]. These tests consist in mortar prisms exposed to solutions containing 4 and 44 g/L of sodium sulfate. The estimated initial composition of the main hydrated products of the mortar is listed in Table 8.1 [16]. The C-S-H phase was modeled as a composite of jennite ($C_{1.7}SH_{2.1}$), tobermorite ($C_{0.8}SH_{1.3}$), and SiO_2, considered as three separate solids in the reactive transport calculations as solid solution models are not yet available in ALLIANCES. The volume fraction of sand is 0.58 and the initial porosity is 0.151.

The mesh of one quarter of the specimen is composed of 50 × 50 identical elements and the simulations are performed in 2D. Note that to reproduce the periodic changes of solution at 7, 14, 28, and 56 days which lead to significant temporal evolutions of ionic species at the specimen surface [16], the solution domain is also meshed (see Fig. 8.1).

To meet the conditions relative to the existence of crystallization pressures in (8.2), a kinetic law is introduced to guarantee the supersaturation of the solution regarding the precipitation of ettringite:

$$\frac{\mathrm{d}C_{\mathrm{ettr}}}{\mathrm{d}t} = k_{\mathrm{p}}(\Omega_{\mathrm{ettr}} - 1) \tag{8.5}$$

where C_{ettr} designates the concentration of ettringite, k_{p} is a kinetics coefficient, and $\Omega_{\mathrm{ettr}} = Q_{ettr}/K_{\mathrm{ettr}}$ where Q_{ettr} is defined in (8.2). The effects of k_{p} are illustrated in Fig. 8.2 showing the profiles of portlandite, ettringite, and gypsum for three

Table 8.1 Composition of the mortar [16]

	Portlandite	Jennite	AFm	AFt	Hemicarbo	Hydrotalcite
mol/L of mortar	1.654	1.387	0.135	0.0165	0.0460	0.0411

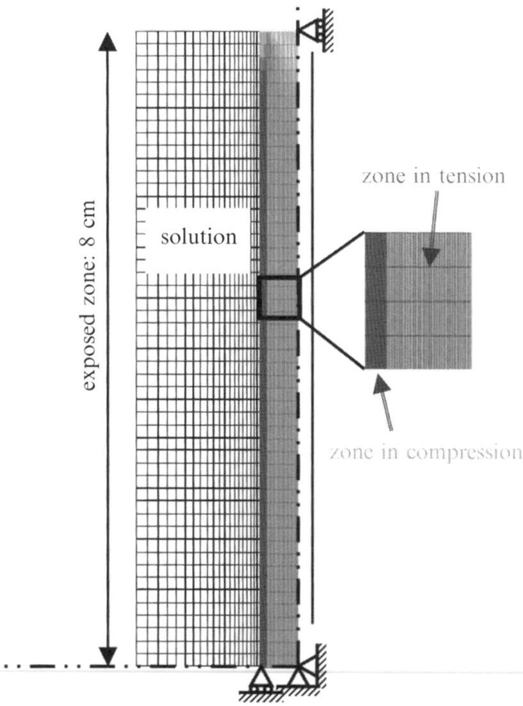

Fig. 8.1 Mesh used for the coupled simulations and vertical stress field in the specimen at 56 days for the 44 g/L sodium sulfate case

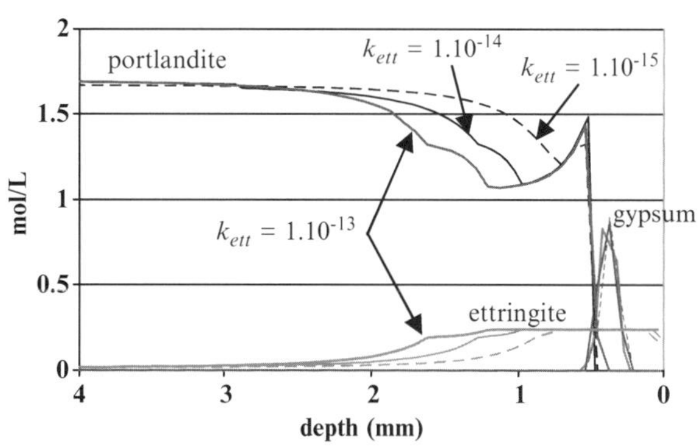

Fig. 8.2 Mineral assemblage profiles obtained at 56 days for the 44 g/L Na_2SO_4 ingressing solution and for three different values of the kinetics constant k_p

different values of the kinetics constant k_p at 56 days for the 44 g/L Na_2SO_4 ingressing solution. We observe that, as expected, ettringite precipitates further in the material for $k_p = 1 \times 10^{-13}$ mol/s than for $k_p = 1 \times 10^{-15}$ mol/s. Portlandite is also modified by the kinetics constant, whereas gypsum appears almost unaffected.

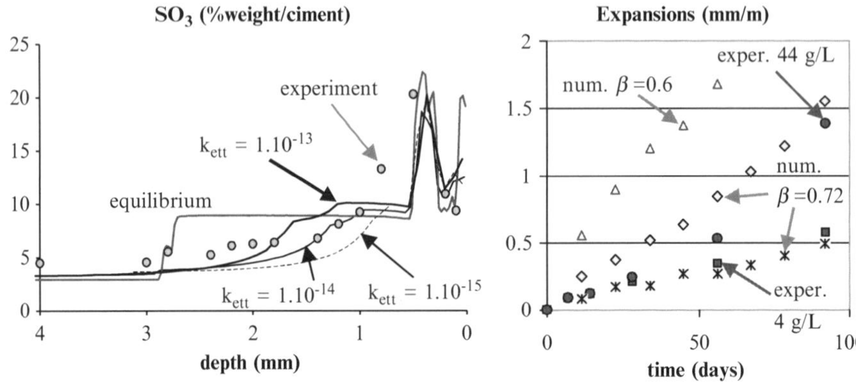

Fig. 8.3 SO_3 profiles as a weight fraction of the mass of cement present at 56 days for the 44 g/L Na_2SO_4 ingressing solution and different values of k_p (*left*), and vertical expansions for both 4 and 44 g/L Na_2SO_4 ingressing solution (*right*)

Figure 8.3 left shows in black lines the calculation results of SO_3 profiles as a weight fraction of the mass of cement present, at 56 days in the 44 g/L Na_2SO_4 case, and for the three aforementioned values of k_p. These profiles are directly related to EDS measurements presented in gray symbols [15]. The profile corresponding to a chemical equilibrium simulation [i.e., without considering (8.5)] is also shown in gray thick line [16]. Figure 8.3 left illustrates the accumulation of sulfate as the sodium sulfate solutions migrates into the mortar. In the core of the sample, the calculated and the measured SO_3 fractions correspond to the total SO_3 present in the cement while they increase nearer to the surface where the minerals interacts with the ingressing Na_2SO_4 solution. We note that the calculations with $k_p = 1 \times 10^{-13}\,\mathrm{mol/s}$ agrees relatively well with the measured weight fractions compared to the other k_p values. However, the simulations indicate generally a narrower zone of precipitated gypsum than the one observed experimentally between about 0.25–1 mm.

Figure 8.3 right presents the time evolutions of the calculated free expansions of the mortar specimens for both 4 and 44 g/L Na_2SO_4 ingressing solution and for two values of the coefficient β introduced in (8.3). These numerical results are compared to experimental data (dark symbols, [15]). We note that the value $\beta = 0.72$ gives results of free expansions close to the measured values for the two considered cases, while $\beta = 0.6$ overestimates them significantly for the 44 g/L Na_2SO_4 case. The approach described in (8.3) and (8.4) appears then to satisfactorily reproduce the observed free expansions, contrary to the crystallization pressure approach alone, which leads to notable underestimations (unless exhibiting unrealistically high crystallization pressures of several thousands of MPa [10]). The same conclusion has been reported in [11]. However, the results obtained in this study do not include the possible microcracking which may be caused by the expansions. In fact, the simulations indicate the presence of relatively high stress levels in both tensile and compressive zones (see Fig. 8.1), suggesting that microcracks may initiate and propagate in particular in the regions subjected to tensile states of stresses.

Moreover, while the approach defined by (8.3) and (8.4) seems appropriate to estimate the free expansions of laboratory specimens, we point out the fact that for real structures or specimens with restrained displacements, the macroscopic stresses resulting from (8.4) may be extremely high. This is due to the expression of the bulk strain $\boldsymbol{\varepsilon}^*$, which does not depend on the stresses. These aspects are currently under investigation, and improvements of the model are expected in particular regarding the calculation of $\boldsymbol{\varepsilon}^*$.

5 Conclusions

We have presented in this study some numerical results regarding the coupled chemo-transport-mechanical simulations of 2D mortar specimens subjected to ESA in saturated conditions. The simulations are performed with the numerical platform ALLIANCES. Crystallization pressures due to restrained ettringite precipitation are assumed to be the major cause of the observed macroscopic expansions. The necessary condition for their existence, i.e., the supersaturation of the solution, is fulfilled by the introduction of a kinetics law governing the ettringite precipitation. To reproduce the free expansions, a macroscopic bulk strain $\boldsymbol{\varepsilon}^*$ calculated from a part of the volume fraction of precipitated ettringite is introduced in the mechanical model. The kinetics constant governing the ettringite precipitation is adjusted so that experimental and numerical mineral concentration profiles of mortar specimens subjected to two different sodium sulfate solutions match reasonably well. The calculated macroscopic free expansions are in quite good agreement with experimental data, provided a correct calibration of the parameter involved in the expression of $\boldsymbol{\varepsilon}^*$. However, it is asserted that the model would lead to very high stress levels in the structures in the particular case of restrained displacements at its boundaries. In particular, we suggest that in the same way that the capillary pressure p_c expressed in (8.2) acts on the solid skeleton to produce macroscopic stresses, external applied (compressive) stresses generate local microscopic stresses in the ettringite inclusions which counteract p_c. As a consequence, formation of ettringite may be reduced and even inhibited by these stresses. This aspect, together with the fact that microcracking is disregarded at this stage, will be the subject of subsequent investigations so as to improve the model response.

Acknowledgments AREVA NC is gratefully acknowledged for financial support.

References

1. Taylor HFW, Famy C, Scrivener KL (2001) Delayed ettringite formation. Cement Concr Res 31:683–693
2. Xie P, Beaudoin JJ (1992) Mechanism of sulfate expansion.1. Thermodynamic principle of crystallization pressure. Cement Concr Res 22:631–640

3. Scherer GW (1999) Crystallization in pores. Cement Concr Res 29:1347–1358
4. Flatt RJ (2002) Salt damage in porous materials: how high supersaturations are generated. J Crys Growth 242:435–454
5. Montarnal P, Mugler C, Colin J, Descostes M, Dimier A, Jacquot E (2007) Presentation and use of a reactive transport code in porous media. Phys Chem Earth A/B/C 32:507–517
6. Bary B, Béjaoui S (2006) Assessment of diffusive and mechanical properties of hardened cement pastes using a multi-coated sphere assemblage model. Cement Concr Res 36:245–258
7. Stora E, Bary B, He QC (2009) Modelling and simulations of the chemo-mechanical behaviour of leached cement-based materials. Leaching process and induced loss of stiffness. Cement Concr Res 39:763–772
8. Stora E (2007) Multi-scale modelling and simulations of the chemo-mechanical behavior of degraded cement-based materials. PhD thesis, University of Marne-La-Vallée, France
9. Bary B (2008) Simplified coupled chemo-mechanical modeling of cement pastes behavior subjected to combined leaching and external sulfate attack. Int J Numer Anal Meth Geomech 32:1791–1816
10. Bary B, Leterrier N, Deville E, Le Bescop P (2010) Coupled chemo-transport-mechanical modeling and simulation of mortar external sulfate attack. In: Proceedings of CONMOD 2010, Lausanne, Switzerland
11. Basista M, Weglewski W (2009) Chemically assisted damage of concrete: a model of expansion under external sulfate attack. Int J Damage Mech 18:155–175
12. Tixier R, Mobasher B (2003) Modeling of damage in cement-based materials subjected to external sulfate attack. I: formulation. J Mater Civ Eng 15:305–313
13. Idiart AE, López CM, Carol I (2011) Chemo-mechanical analysis of concrete cracking and degradation due to external sulfate attack: a meso-scale model. Cement Concr Compos 33:411–423
14. Planel D, Sercombe J, Le Bescop P, Adenot F, Sellier A, Capra B, Torrenti JM (2001) Experimental and numerical study on the effect of sulfates on calcium leaching of cement paste, FRAMCOS, Cachan, France, May 28–June 01, 2001, pp 287–292
15. Schmidt T (2007) Sulfate attack and the role of internal carbonate on the formation of thaumasite. EPFL, Lausanne
16. Lothenbach B, Bary B, Le Bescop P, Schmidt T, Leterrier N (2010) Sulfate ingress in Portland cement. Cement Concr Res 40:1211–1225

Chapter 9
Micro- and Macroscopic Investigations of Actinide Binding in Cementitious Materials

E. Wieland, R. Dähn, X. Gaona, N. Macé, and J. Tits

1 Introduction

The Swiss radioactive waste disposal program, as many others worldwide, foresees the long-term safe disposal of low- and intermediate-level radioactive waste (L/ILW) in a cement-based deep geological repository. In the proposed Swiss concept it is foreseen that cement is used to condition the waste materials and for the construction of the engineered barrier (liner, backfill material). Hence, cementitious materials are an important component of the multi-barrier system in underground repositories. In performance assessment it is considered that the cementitious near field controls radionuclide solubility and retards radionuclide release into the far field. Therefore, knowledge of the chemical mechanisms by which radionuclides are retarded due to interaction with cement minerals is essential for long-term assessments of the safe disposal of radioactive waste.

Hardened cement paste (HCP) is a heterogeneous matrix consisting of mainly calcium (aluminium) silicate hydrates [C-(A)-S-H], portlandite [$Ca(OH)_2$] and calcium aluminates (AFt- and AFm-type phases). Among the different cement minerals, C-S-H phases are considered to be the most important constituent of HCP as they are abundant and exhibit a wide diversity of structural sites for cation and anion binding.

Actinides are common constituents in L/ILW. Uranium(VI) is expected to be the dominant redox state under the relevant repository conditions, i.e. slightly oxidizing to reducing conditions. Several studies have been published on U(VI) uptake by HCP and cement minerals (e.g. [1–5]).

E. Wieland (✉) • R. Dähn • N. Macé • J. Tits
Laboratory for Waste Management, Paul Scherrer Institute, Villigen, Switzerland
e-mail: erich.wieland@psi.ch

X. Gaona
Karlsruhe Institute of Technology, Institut für Nukleare Entsorgung, Karlsruhe, Germany

F. Bart et al. (eds.), *Cement-Based Materials for Nuclear Waste Storage*,
DOI 10.1007/978-1-4614-3445-0_9,

Spectroscopic studies on the sorption mechanism of U(VI) in cementitious materials complement our current understanding from the batch sorption experiments [6–10]. Extended X-ray absorption fine structure (EXAFS) spectroscopy studies of Sylvester et al. [7] and Zhao et al. [10] showed that the uranyl (UO_2^{2+}) moiety is preserved on U(VI) loaded untreated and hydrothermally treated cement samples, indicating that the +VI redox state is stable upon uptake by HCP. These studies further showed that U(VI) equilibrated with hydrothermally altered concrete formed oligomeric surface complexes or surface precipitates. The authors proposed that U(VI) coordination onto the surfaces of cement minerals occurs via equatorial oxygen bonding. The results reported by Harfouche et al. [6], Wieland et al. [9] and Tits et al. [8] will be discussed in the following section.

In contrast to its importance for performance assessment due its long half-life (^{237}Np, $t_{1/2} = 2.14 \times 10^6$ years), its radiotoxicity and its redox sensitivity, the currently available information on the behaviour of Np in cementitious environments is very limited. Np(IV) species are expected to exist under the reducing conditions of a repository near field. Zhao et al. [10] observed partial reduction of Np(V) to Np(IV) in the course of Np(V) uptake by HCP while the reduction process had no influence on the overall partitioning between the aqueous phase and HCP, indicating similar sorption behaviour. At present, sorption studies with Np(IV) on cementitious materials are lacking. On the basis of chemical analogy with Th(IV), Np(IV) is expected to be strongly retarded by HCP and C-S-H while a significantly weaker uptake of Np(V) by HCP is expected due to its lower effective charge (+2.3). Note, however, that preliminary in-house measurements suggest strong interaction with cementitious materials, comparable to that of tetravalent actinides.

In this chapter, we report results from micro-spectroscopic investigations (μ-XRF(X-ray fluorescence)/EXAFS/XANES) (X-ray absorption near edge spectroscopy) on U(VI) and Np(V,VI) loaded intact HCP samples, which are discussed in connection with bulk EXAFS studies on U(VI) and Np(IV,V,VI) loaded HCP and C-S-H samples.

2 Material and Methods

2.1 Sample Preparation

Intact samples. U(VI) and Np(V,VI) loaded intact HCP samples were prepared using a sulphate resistant Portland cement denoted as HTS cement [CEM I 52.5 N HTS (=Haute Teneur en Silice), Lafarge, France] [11]. The U(VI) and Np(V,VI) loaded hydration samples were prepared by mixing unhydrated HTS cement with U(VI) and Np(V) stock solutions of appropriate concentrations at a water:cement (w/c) ratio of 0.4 following the procedure described in Vespa et al. [12]. Mass balance calculations showed that, after addition of the actinide stock solutions, the conditions in the pastes remained strongly alkaline due to the large buffering capacity of cement. Sodium hypochlorite (NaOCl) 5×10^{-3} M was added to one of the hydrating samples to promote the oxidation of Np(V) to Np(VI). The average U(VI) loading of the cement paste was ~20,000 ppm in the U(VI) sample while the Np loading was ~1,000 ppm in the Np(V,VI) samples. The U(VI) and Np(V,VI) loaded pastes were cast in Plexiglass moulds, which were tightly closed using polyethylene snap caps. The samples were stored at room temperature and 100 % relative humidity in a closed container, which was placed in the glovebox with N_2 atmosphere, and cured over appropriate periods of time [U(VI): 1 day; Np(V,VI): 30 days]. Upon curing the HCP samples were extruded from the cast, immersed in acetone for 12 h to stop the hydration process and dried overnight in the glovebox. From the HCP material, polished thin sections were prepared for further use in the μ-XRF/XAS investigations.

Sorption samples. The actinide doped HCP and C-S-H samples were prepared according to procedures described in detail elsewhere [13, 14]. Briefly, U(VI) and Np(IV,V,VI) stock solutions were added to synthetic C-S-H suspensions [C-S-H with various Ca/Si (C/S) ratios] or crushed HCP material in equilibrium with artificial cement pore water. The Np(IV) stock solution was prepared in a glovebox by electrolysis of Np(V) while NaOCl was added to oxidize Np(V) to Np(VI) [13]. Sodium dithionite was added to stabilize Np(IV). The pH of the actinide doped sorption samples ranged from 10 to 13.3 and the loadings from ~860 ppm [Np(IV,V, VI)] to ~2,700 ppm [U(VI)]. After equilibration and separation of the solid and liquid phases, the wet pastes were placed into Plexiglass holders, sealed with two layers of Kapton® tape and, in addition, with thermo-sealed polyethylene bags. Note that special precautions had to be taken during preparation and transport of the Np(IV) loaded samples to the beamline in order to avoid/minimize access of oxygen [13]. Reference samples were prepared for the U(VI) studies by diluting ~15 mg of the U(VI) minerals with 250 mg boron nitride (BN, 98 % purity, Sigma-Aldrich, Switzerland). A description of the U(VI) minerals (Ca uranate, uranophane, soddyite) is given elsewhere [14]. No Np reference compounds were available.

2.2 Spectroscopic Measurements

U(VI) micro-spectroscopic measurements. Micro-XRF/XAS measurements were carried out at BL 2-3 at the Stanford Synchrotron Radiation Lightsource (SSRL, Menlo Park, CA, USA). The beamline is equipped with a Si(111) double-crystal monochromator and a Kirkpatrick–Baez (KB) mirror focusing system. The monochromator position was calibrated by assigning the first inflection point of the K-absorption edge of a Y foil to 17,038 eV. The X-ray beam was focussed down to ~2×2 μm^2 spot size. Fluorescence radiation was detected using a single channel VORTEX Si detector. Several EXAFS spectra were recorded on the selected U(VI)-rich spot to achieve the requested signal-to-noise (S/N) ratio.

Np(V,VI) micro-spectroscopic measurements. Micro-XRF/XANES was carried out at the MicroXAS beamline at the Swiss Light Source (SLS, Villigen PSI, Switzerland). Monochromator and focusing system correspond to that at BL 2-3, and the same adjustments were made (energy calibration with a Y foil; ~2 × 2 μm^2 spot size). Fluorescence radiation was detected using a KETEK Si detector. Several XANES spectra were recorded to achieve the requested S/N ratio.

Bulk EXAFS measurements. EXAFS spectra at the U L_{III} edge (17,166 eV) and the Np L_{III} edge (17,610 eV) were collected at room and liquid He temperatures, respectively, at BM 20 (ROBL) at the European Synchrotron Radiation Facility (ESRF, Grenoble, France). Description of the beamline and the experimental set-ups are described elsewhere [13, 14].

Data analysis. Bulk-EXAFS and μ-EXAFS data reduction and analysis were performed using the SIXPACK and IFEFFIT software packages following standard procedures [15–17]. Reduction and modelling of the EXAFS data were performed with the ATHENA/ARTEMIS software package (for details see [13, 14]).

Analysis of the EXAFS spectra was based on the assumption that several species of the absorber atom contributed to the overall EXAFS signal. Therefore, the EXAFS spectra were analysed on the basis of statistical analysis using principal component analysis (PCA) and iterative target-transformation factor analysis (ITFA) as described elsewhere (e.g. [18, 19]). PCA was used to determine if a set of spectra can be represented as linear combination of a smaller number of independent component spectra. The indicator value as described in Manceau et al. [18] was used as the criterion to determine the number of independent components sufficient to reproduce the set of experimental spectra. Linear combination (LC) fitting was carried out with the aim of describing the original spectra as a sum of the new components with coefficients (weights) between 0 and 1. Structural information was obtained by applying a multi-shell approach for EXAFS data fitting of those components identified by PCA [13, 14].

3 Results and Discussion

3.1 U(VI) Speciation in Cementitious Systems

The findings from the bulk EXAFS and TRLFS investigations on U(VI) loaded HCP and C-S-H samples can be summarized as follows [6, 8, 9, 14].

Sorbed U(VI) species. TRLFS allowed the determination of two different sorbed species along with a U(VI) precipitate (Ca uranate) and the aqueous U(VI) hydroxo species present under alkaline conditions [8]. The spectral characteristics of the sorbed species indicated strong coordination via equatorial oxygen atoms, suggesting inner-sphere complexation. The sorbed species with the weaker spectral red shift is expected to be a surface-bound species or a species weakly bound in the

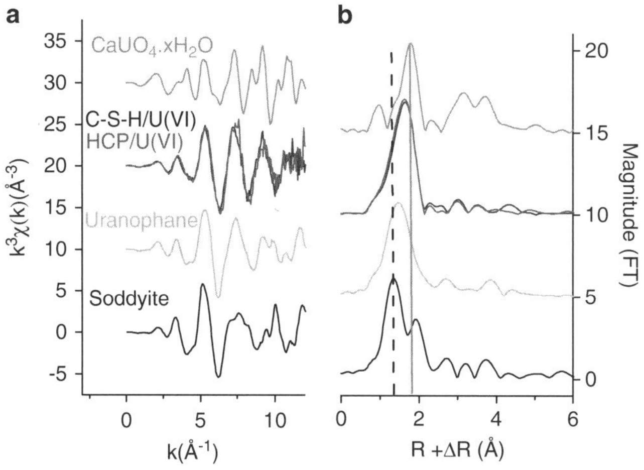

Fig. 9.1 (**a**) k^3-weighted EXAFS data for selected U(VI) reference compounds and U(VI)/HCP and U(VI)/C-S-H sorption samples; (**b**) the corresponding radial structure function with the k range taken between 2 and 11 $Å^{-1}$ for all Fourier transforms

interlayer of C-S-H. The proportion of this species was found to decrease over time. The species with the stronger red shift is considered to be incorporated in the C-S-H interlayer. The proportion of this species increased over time. The presence of an incorporated U(VI) species was already suggested based on an earlier EXAFS study [6].

C-S-H as uptake-controlling cement phase. TRLFS and EXAFS studies further showed that the coordination environment of U(VI) is similar upon uptake by HCP and C-S-H phases [8, 14]. From this finding we infer that C-S-H is the uptake-controlling cement phase in the complex cement matrix. Micro-spectroscopic investigations on U(VI) loaded intact HCP samples, whereby U(VI) was allowed to diffuse into the cement matrix over a period of 270 days, showed that U(VI) preferentially accumulates in the reactive zones around Ca-rich clinker minerals, presumably belite (Ca_2SiO_4), where C-S-H phases form [9]. Thus, U(VI) speciation was found to be identical regardless of whether U(VI) uptake starts from over-saturation as in the sorption experiments due to the addition of a highly concentrated U(VI) stock solution or from under-saturation as in the diffusion experiments where the U(VI) concentration was fixed at the solubility limit of Ca uranate [9].

U(VI) coordination environment. U(VI) taken up by C-S-H phases has a local arrangement of the nearest neighbouring atoms that resembles the coordination environment of U(VI) in the uranyl silicates uranophane and soddyite. While EXAFS indicated formation of an uranophane-like structure (Fig. 9.1), TRLFS suggested a soddyite-like structure on the basis of the position (red shift) of the emission bands in the fluorescence spectra [8]. At the present time the reason for this discrepancy is not understood. Nevertheless, irrespective of the discrepancy,

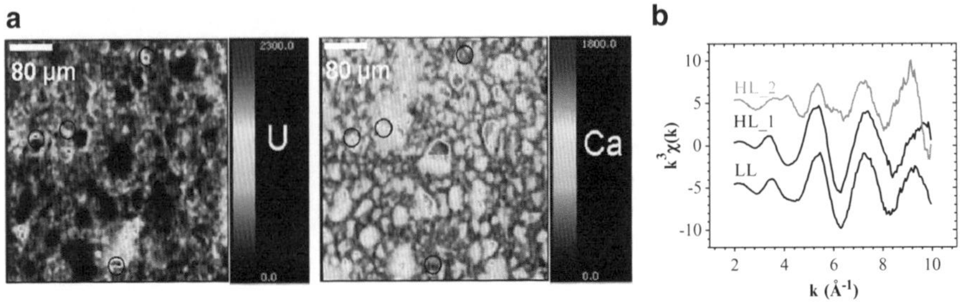

Fig. 9.2 (**a**) Micro-XRF maps of U and Ca distribution in a U(VI) loaded HCP sample [~20,000 ppm U(VI), 1 day hydration]; µ-EXAFS spectra were collected on the indicated U-rich spots (*circles*). (**b**) k^3-weighted EXAFS spectra of the main components obtained with PCA. All composite spectra collected at the U spots contain two U(VI) species designated as HL_1 and HL_2 species whereby the LL and HL_1 species are identical. *K*-range for data analysis was taken between 2 and 10 $Å^{-1}$

we may conclude that U(VI) is taken up into the structure of C-S-H phases, where it occupies crystallographic positions similar to those of U(VI) in uranyl silicates. This further supports the idea that, after long equilibration times, U(VI) preferentially forms an incorporated species in the C-S-H structure.

U(VI) precipitate. At high U(VI) loadings (20,000 ppm) U(VI) forms a Ca uranate in cementitious materials. Precipitation of Ca uranate was previously postulated from thermodynamic modelling of U(VI) solubility limits in C-S-H [5] and cement systems [9], and further supported by TRLFS measurements on U(VI) loaded sorption samples [8]. Figure 9.2 shows that, in intact U(VI) loaded HCP samples with high U(VI) loadings, U(VI) is heterogeneously distributed in the cement matrix and further, the EXAFS spectra collected at U(VI) hot spots contain backscattering contributions from two species (a) the "U(VI)-C-S-H-type" species (as described earlier) and (b) Ca uranate. Hence, Ca uranate is considered to be the solubility-limiting phase in cementitious systems in short-term experiments (up to 300 days). Whether this phase is stable in the long term, however, is currently uncertain. Kienzler et al. [20] suggested that uranophane could be the solubility-limiting U(VI) mineral in the long term, particularly in cement matrices corroded by NaCl and $MgCl_2$ brines.

3.2 Np Speciation in Cementitious Materials

Information on the speciation of Np in cementitious materials is very limited and mainly results from a recent bulk EXAFS study on Np loaded HCP and C-S-H systems [13] and an ongoing micro-spectroscopic study.

Sorbed Np species. The results from the bulk EXAFS investigation on Np(IV) loaded HCP and C-S-H samples indicated predominant incorporation of Np in the C-S-H structure for C/S ratios ranging between 0.75 and 1.65. Two species were

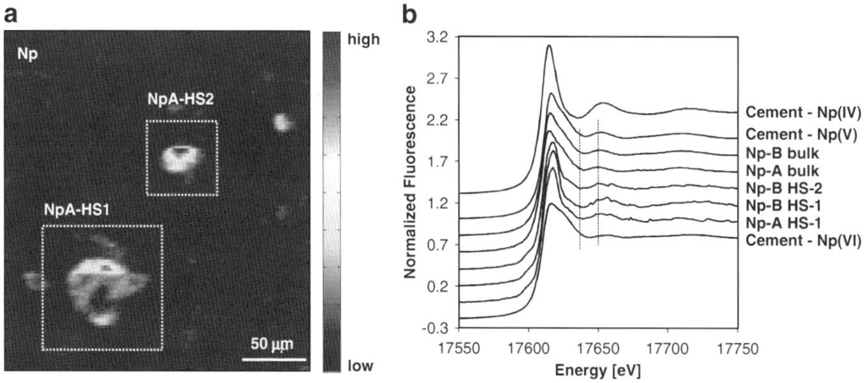

Fig. 9.3 Micro-XRF maps of Np collected on a Np(V) loaded HCP sample after 30 days of hydration (**a**). Np-rich spots are denoted as NpA-HS1/2. (**b**) Comparison of XANES (*bulk*) and μ-XANES (*hot spots*) spectra determined on the two samples: A = Np(V) loaded HCP (**a**), B = Np(V) tracer + NaOCl loaded HCP (map not shown). Bulk-XANES data of the Np(IV,V,VI) loaded HCP sorption samples are given as references

identified, corresponding to Np(IV) in C-S-H with C/S = 1.65 as in "fresh" HCP and 0.75 as in highly degraded HCP. The coordination environment of Np(IV) was found to depend on the C/S ratio of C-S-H but not on the pH in accordance with the aqueous speciation of Np(IV) in the pH range 10–13.3 [predominance of Np $(OH)_4(aq)$]. Comparison of the structural parameters determined with EXAFS on the Np(IV)/HCP and Np(IV)/C-S-H samples further showed the same coordination environments of Np(IV), indicating that C-S-H are responsible for Np(IV) immobilization in cementitious materials. Hence, as for U(VI), incorporation of Np(IV) by C-S-H phases determines the long-term immobilization of this tetravalent actinide in the cementitious near field. Preliminary analysis of EXAFS data further suggests that incorporation into the structure of C-S-H phases also could play an important role in the immobilization of pentavalent neptunium.

Np speciation in intact HCP samples. Np was found to be heterogeneously distributed in the cement matrix (Fig. 9.3a). Preferential accumulation was observed around reacting clinker minerals as in the case of U(VI) [9]. Micro-XANES further showed that the redox state of Np in the samples in presence of NaOCl does not change (Fig. 9.3b). In contrast to observations made on Np sorption samples using HCP, NaOCl did not oxidize Np(V) to Np(VI) within the time scale of the hydration process. One may speculate that Np(V) was taken up rapidly by C-S-H phases during cement hydration, which impeded oxidation by NaOCl. Comparison of μ- and bulk-XANES data further revealed differences in the coordination environments of Np(V) in the bulk material and on the local spots. XANES indicates presence of both Np(IV,V) oxidation states at spots NpA-HS1 and NpB-HS1/2 as the broad shoulder, which is characteristic for Np(V), is absent in the spectra (Fig. 9.3b). The reason for the presence of Np(IV) is presently unknown although beam damage could be a plausible explanation.

4 Conclusion

Macro- and micro-spectroscopic investigations provide essential information on the uptake mechanisms of actinides by cementitious materials. This information is complementary to that obtained from batch sorption studies and discloses new aspects in conjunction with the long-term safe disposal of actinides. The spectroscopic studies indicate that, in addition to the formation of surface-bound species, actinides can be readily taken up into the interlayer of the C-S-H structure where they occupy specific crystallographic positions [e.g. as in uranyl silicates in the case of U(VI)]. Furthermore, actinides are preferentially accommodated by C-S-H phases in the cement matrix. This finding, in combination with the observed long-term persistence of C-S-H phases in an evolving cementitious near field, further implies that the uptake-controlling cement phase for actinides might be present in a cement-based repository over a very long period of time. This strongly suggests that long-term safe disposal of actinides in a cementitious near field is possible.

Acknowledgements Portions of this research were carried out at the Stanford Synchrotron Radiation Lightsource, a directorate of SLAC National Accelerator Laboratory and an Office of Science User Facility operated for the U.S. Department of Energy Office of Science by Stanford University. The SSRL Structural Molecular Biology Program is supported by the DOE Office of Biological and Environmental Research, and by the National Institutes of Health, National Center for Research Resources, Biomedical Technology Program (P41PR001209). The SLS (Villigen PSI, Switzerland) and ESRF (Grenoble, France) are acknowledged for the provision of beam time. The authors thank the beamline scientists of the following facilities for their technical support: MicroXAS at SLS, BL 2-3 at SSRL and BM20 (ROBL) at ESRF. Thanks are extended to Dr. M. Marques-Fernandes and D. Kunz for assistance during measuring campaigns. Prof F.P. Glasser is gratefully acknowledged for provision of the uranium reference materials. Partial financial support was provided by the "ACTINET" project [sixth EC framework programme (FP)], by the European Project "RECOSY" (seventh EC FP) and by the National Cooperative for the Disposal of Radioactive Waste (Nagra), Switzerland. X.G. acknowledges the Generalitat de Catalunya for his post-doctoral grant ("Beatriu de Pinós" program). Portions of this research were carried out in the framework of a Marie Curie Fellowship with a grant to N.M. financed by the EC (Contract No. FP6-044811).

References

1. Bayliss S, McCrohon R, Oliver P, Pilkington NJ, Thomason HP (1996) Near field sorption studies: Jan 1989 to June 1991, NSS/R227, AEA-ESD-0353
2. Brownsword M, Buchan AB, Ewart FT, McCrohon R, Ormerod GJ, Smith-Briggs JL, Thomason HP (1990) The solubility and sorption of uranium(VI) in a cementitious repository. Mater Res Soc Symp Proc 176:577–582
3. Pointeau I, Landesman C, Giffault E, Reiller P (2004) Reproducibility of the uptake of U(VI) onto degraded cement pastes and calcium silicate hydrates. Radiochim Acta 92:645–650
4. Sutton M, Warwick P, Hall A (2003) Uranium(VI) interactions with OPC/PFA grout. J Environ Monit 5:922–925
5. Tits J, Fujita T, Tsukamoto M, Wieland E (2008) Uranium(VI) uptake by synthetic calcium silicate hydrates. Mater Res Soc Symp Proc 1107:467–474

6. Harfouche M, Wieland E, Dähn R, Fujita T, Tits J, Kunz D, Tsukamoto M (2006) EXAFS study of U(VI) uptake by calcium silicate hydrates. J Colloid Interface Sci 303:195–204
7. Sylvester ER, Allen PG, Zhao P, Viani BE (2000) Interactions of uranium and neptunium with cementitious materials studied by XAFS. Mater Res Soc Symp Proc 608:307–312
8. Tits J, Geipel G, Macé N, Eilzer M, Wieland E (2011) Determination of uranium(VI) sorbed species in calcium silicate hydrate phases: a laser-induced luminescence spectroscopy and batch sorption study. J Colloid Interface Sci 359:248–256
9. Wieland E, Macé N, Dähn R, Kunz D, Tits J (2010) Macro- and micro-scale investigations on U(VI) immobilization in hardened cement paste. J Radioanal Nucl Chem 286:793–800
10. Zhao P, Allen PG, Sylvester ER, Viani BE (2000) The partitioning of uranium and neptunium onto hydrothermally altered concrete. Radiochim Acta 66:729–736
11. Lothenbach B, Wieland E (2006) A thermodynamic approach to the hydration of sulphate-resisting Portland cement. Waste Manage 26(7):706–719
12. Vespa M, Dähn R, Grolimund D, Wieland E, Scheidegger AM (2006) Spectroscopic investigation of Ni speciation in hardened cement paste. Environ Sci Technol 40(7):2275–2282
13. Gaona X, Dähn R, Tits J, Scheinost AC, Wieland E (2011) Uptake of Np(IV) by C-S-H phases and cement paste: an EXAFS study. Environ Sci Technol 45:8765–8771
14. Macé N, Dähn R, Tits J, Scheinost AC, Wieland E (2011) EXAFS investigation on U(VI) immobilization in hardened cement paste: influence of experimental conditions on speciation. Radiochim Acta (submitted)
15. Newville M (2001) EXAFS analysis using FEFF and FEFFIT. J Synchrotron Radiation 8:96–100
16. Ravel B, Newville M (2005) ATHENA, ARTEMIS, HEPHAESTUS: data analysis for X-ray absorption spectroscopy using IFEFFIT. J Synchrotron Radiation 12:537–541
17. Webb SM (2005) Sixpack: a graphical user interface for XAS analysis using IFEFFIT. Phys Scr T115:1011–1014
18. Manceau A, Marcus MA, Tamura N (2002) Quantitative speciation of heavy metals in soils and sediments by synchrotron X-ray techniques. In: Fenter P, Sturchio NC (eds) Applications of synchrotron radiation in low-temperature. Geochemistry and environmental science, vol 49, Reviews in Mineralogy and Geochemistry. Mineralogical Society of America, Washington, DC, pp 341–428
19. Rossberg A, Reich T, Bernhard G (2003) Complexation of uranium(VI) with protocatechuic acid-application of iterative transformation factor analysis to EXAFS spectroscopy. Anal Bioanal Chem 376:631–638
20. Kienzler B, Metz V, Brendebach B, Finck N, Plaschke M, Rabung Th, Rothe J, Schild D (2010) Chemical status of U(VI) in cemented waste forms under saline conditions. Radiochim Acta 98:675–684

Chapter 10
A New Radionuclide Sorption Database for Benchmark Cement Accounting for Geochemical Evolution of Cement

L. Wang, M. Ochs, D. Mallants, L. Vielle-Petit, E. Martens, D. Jacques, P. De Cannière, J.A. Berry, and B. Leterme

1 Introduction

ONDRAF/NIRAS is developing a safety case as part of a license application for building and operating a near-surface disposal facility for short-lived low and intermediate level waste (LILW-SL) in Dessel, Belgium. Safety assessment (SA) is being performed by SCK•CEN to underpin the safety case. In the ONDRAF/NIRAS disposal concept, cementitious materials are used as the major physical and chemical barriers to the release of radionuclides from the disposal facility (see Fig. 10.1 for the design concept). Chemical containment or immobilisation resulting in retardation of contaminant migration is expected to be realised mainly by sorption on concrete engineered barriers. For the current SA calculations, sorption of radionuclides is accounted for by a linear and reversible sorption process. Sorption values in the form of distribution coefficients (K_d, [L/kg]) are therefore needed. In sorption literature, however, the majority of data are reported as distribution ratios (R_d, [L/kg]) without clear knowledge if the value was determined under conditions of linear, reversible and equilibrium sorption. For SA purposes, the selected sorption values must be applicable to conditions suitable for the use of K_d values. Concerning terminology in this chapter, sorption refers to surface sorption (in other words, no distinction is made between surface adsorption, incorporation, binding by lattice substitution, etc.) but excluding bulk precipitation.

L. Wang (✉) • D. Mallants • E. Martens • D. Jacques • P. De Cannière • B. Leterme
Institute for Environment, Health, and Safety, Belgian Nuclear Research Centre (SCK•CEN), 2400 Mol, Belgium
e-mail: lwang@sckcen.be

M. Ochs • L. Vielle-Petit
BMG Engineering Ltd., Ifangstrasse 11, CH 8952 Schlieren, Switzerland

J.A. Berry
Serco TCS, Harwell, Didcot OX11 0QB, United Kingdom

F. Bart et al. (eds.), *Cement-Based Materials for Nuclear Waste Storage*,
DOI 10.1007/978-1-4614-3445-0_10,

Fig. 10.1 The ONDRAF/NIRAS near-surface disposal concept for LILW-SL disposal in Dessel, Belgium [1]

Compilations of sorption values for cement available in the open literature, e.g. Bradbury and Sarott [2], Bradbury and Van Loon [3], Wieland and Van Loon [4], Allard [5], and Baker et al. [6] were used in a previous iteration of the SA [7]. A new data compilation was needed for the current SA calculations for the following reasons:

- Since the previous SA iteration, more detailed geochemical boundary conditions for the cementitious components of the Dessel repository have become available. Therefore, sorption values need to be re-evaluated for the more relevant and site-specific chemical conditions.
- The majority of the existing sorption values from the literature deal with situations for deep geological disposal where geochemical and hydrogeological conditions are different from the near-surface disposal site at Dessel. For deep disposal, redox conditions are likely to be reducing, while for near-surface disposal, as is proposed here, overall conditions are expected to be close to oxidising. Also, slow groundwater flow in a deep-disposal situation leads to an extremely low rate of cement leaching and thus geochemical evolution, resulting in sorption values that remain relatively constant over time. Because of this, most existing data compilations only need to consider sorption values for cement that has encountered limited geochemical evolution. For the Dessel repository, the rate of water infiltration is expected to be high enough for cement degradation to proceed much faster than under deep disposal conditions. As a result, sorption values for highly evolved cementitious materials are also needed.
- New sorption data have been available since the previous SA iteration, e.g. Pointeau et al. [8], Garcia-Gutierrez et al. [9], and many others. It is necessary to include these state-of-the-art data.
- Also, and distinct from the majority of existing data compilations, an international panel of experts has been involved in the process of data review, selection, and documentation of scientific arguments.

This chapter presents the new sorption database and the procedure of data selection based mainly on literature review and the guidance of an international panel of experts. Details of the data review are given in Wang et al. [10] and Ochs et al. [11].

2 Method for Data Selection

Literature sorption values in the form of K_d or R_d on cementitious materials for 25 elements (Ag, Am, Be, C, Ca, Cl, Cs, H, I, Mo, Nb, Ni, Np, Pa, Pb, Pd, Pu, Ra, Sr, Se, Sn, Tc, Th, U, Zr) were reviewed. Under the guidance of an international panel of experts, best estimate, upper and lower bound values for each element were selected. All sorption values were determined for four degradation states of cementitious materials under the expected local conditions of the Dessel repository. To account for uncertainties about the repository redox conditions, values for both oxidising and reducing conditions were determined for redox sensitive elements. Sorption values for a benchmark cement, which is pure cement without effects of organics, high chloride or other ions, were reviewed and selected. The influence of perturbing components such as organics from waste and cement additives, and other ions was assessed separately.

2.1 Geochemical Conditions Applied to Selection of Sorption Values: Four States of Cement Degradation

Cementitious materials are not thermodynamically stable when in contact with infiltrating water and tend to degrade over time. Cement degradation arises from a number of processes but a principal chemical degradation mechanism expected for the Dessel repository is leaching of cement components by infiltrating soil water. Relevant chemical conditions were determined by reaction modelling assuming equilibrium between cement phases and infiltrating water [12, 13]. Figure 10.2 shows the modelled four states of cement degradation characterised by the evolution of pH, the total Ca and Si solid phases, and the calcium/silicium ratio (C/S) ratio that prevail in the near field.

At the reference temperature of 25 °C, state I has a pH above 13 with a high concentration of Na and K ions in the cement pore fluid. State II has a fixed pH of 12.5 controlled by the solubility of portlandite. State III starts when all portlandite has been leached out of the cement and the pH drops to around 10 when the pore water composition is buffered by C–S–H phases. State IV is free of effective cement phases and has a pH lower than 10 with calcite and aggregate minerals controlling the final pH, together with the chemical composition of the infiltrating soil water. Note that at the ambient temperature of the Dessel repository, estimated to be around 10 °C, the pH will be almost 0.5 units higher than at 25 °C, i.e. 14 for state I, 13 for state II, and so on.

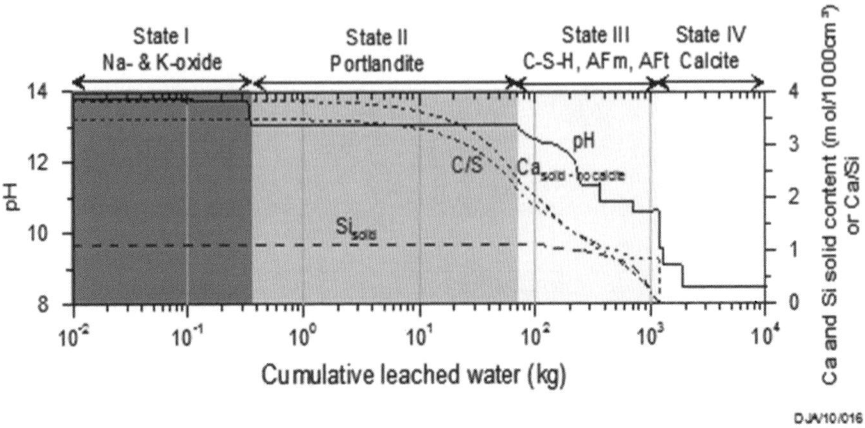

Fig. 10.2 Calculated evolution of pH (*left axis*), solid phase Si_{solid} and $Ca_{solid\text{-}no\ calcite}$ (calcareous aggregates not accounted for) and calcium/silica (C/S) ratio (*right axis*) at 10 °C. Cumulative leached water corresponds to total amount of external (soil) water considered in equilibrium reactions with 1 dm^3 of concrete

2.2 *Procedure for Data Selection*

Sorption values were reviewed from original research papers and reports and presented as K_d or R_d as a function of pH. The procedure for evaluating the data quality and appropriateness is based on the criteria below:

1. Do the sorption data represent K_d or R_d?
2. Are the data from batch test or derived from leaching tests? Preference was given to batch test data while leaching tests were considered to produce sorption values with higher uncertainty owing to use of interpretative models which typically need uncertain model parameters.
3. If modelling was included to interpret experimental data, additional credit was given to such sorption values.
4. For redox sensitive elements: were experiments carried out while keeping control of the oxidation state? In the present study best estimates of sorption values will not be linked with specific values of redox potential (E_h), but instead the different oxidation states of the redox-sensitive radionuclides will be addressed.
5. How was the dominant aqueous speciation of the sorbing element determined? This is normally achieved based on common knowledge of solution chemistry or speciation calculation using thermodynamic models. Only in a very few cases has the speciation of the sorbing element been experimentally determined.
6. What was the liquid-to-solid separation technique? A total separation is important in many cases, especially for strongly sorbing radionuclides.

7. What was the solid-to-liquid ratio (S/L)? As far as equilibrium sorption is concerned, the S/L ratio should normally not influence the sorption value. Experimentally observed variation in sorption values as a function of the S/L ratio may indicate some experimental artefacts or the influence of uptake processes other than sorption.
8. Sorption isotherms were preferred over a single point measurement as they help to evaluate the linearity of the uptake process over a wider range of initial radionuclide concentrations.
9. Was the initial tracer concentration well below its solubility limit to avoid bulk precipitation of the sorbing element?
10. Was the atmosphere in the glove box controlled by means of N_2, Ar, or CO_2? This helps with judging the speciation of the sorbing element and the stability of the cement materials as sorbent, especially when in combination with the addition of a reducing agent.
11. What was the contact time for batch tests? Differences in final sorption values may occur as a result of different contact times, resulting in different degrees of equilibrium. Very long contact times may result in a transition from reversible to irreversible processes, the latter involving slow reactions, e.g. diffusion/incorporation processes. Note that the fact that a concentration does not change with time in a sorption experiment is not generally accepted as a proof of equilibrium. A more generally accepted criterion is reversibility: if the same data are obtained from adsorption and desorption tests, it can be assumed that equilibrium has been attained. However, such information is not always available, and therefore recourse to the former criterion is acceptable. Another reason to prefer the reversibility criterion is its implication when sorption values are used for safety assessments. Use of short-term reversible test results is conservative, in terms of safety, compared to long-term possibly irreversible test values.
12. To which state of cement degradation did the pore water composition correspond?
13. Final pH of the test solution provides an important indication of whether the pore water was in chemical equilibrium with the sorbing solid or if significant interactions occurred between the sorbing solids and the liquid phase.
14. What was the sorbing solid?

These criteria were implemented in an ACCESS™ database. Further guidance was provided by the panel of experts for the literature review and included the following:

1. If sufficient data are available, sorption values of redox-sensitive elements should be included for both oxidising and reducing conditions.
2. Data should focus on cement degradation states II and III where sorption is effectively dominated by cement phases. For state III, values will be taken at about pH 11 (state II has a fixed pH of 12.5 at 25 °C).

3. Sorption data obtained from crushed concrete and mortar are not included because the crushing may generate sorbing surfaces from aggregates.
4. Whenever possible, sorption values on amorphous calcium silicate hydrates (CSH) are preferred because measurements on crystalline CSH phases are not representative of real cements.
5. Sorption on amorphous or poorly crystalline materials, including CSH, may not give a single value but rather a range of 1–2 orders of magnitude.
6. Sorption values determined from experiments involving incompatible substances (resulting in a chemical reaction in the course of the sorption experiment) should be excluded.
7. Sorption values were discussed for fully hydrated cement paste and the corresponding hydrated mineral phases. When there is clear evidence of the presence of unhydrated clinker phases, the data should be excluded.

Temperature influence was not considered because tests are usually done at ambient temperature around 25 °C. Although the temperature in the disposal facility will be lower at ~10 °C, no significant differences are expected in sorption behaviour. Note however that geochemical calculations of cement evolution were done at 10 °C. There is about a 0.5 pH increase in each state compared to such calculations at 25 °C (Jacques 2008).

If no sufficient data were found, supplemental values are estimated based on chemical analogy or other arguments. The purpose is to provide a minimum of guidance to safety assessors in developing data for assessment purposes in the absence of best estimates. The supplemental values are indicative data usually supported by qualitative or semi-quantitative arguments only, and thus are associated with a larger uncertainty.

2.3 Role of the Expert Panel

The expert panel includes representatives of several European countries. This is valuable because their collective experience affords insights into different national programmes and has enabled data which might otherwise have remained obscure to be included in the scope of the review.

The role of the expert panel was to provide guidance on the initial data review carried out by SCK•CEN during panel meetings and endorse the final sorption values. The guidance was given following the check list below:

- *Reliability*: the panel agreed that they must be fully aware of how reliable the data are that they are using. Several issues were discussed:
  - It was generally agreed that data from batch experiments are preferable: data from through-flow experiments are complementary.

  - For traceability and defensibility it was essential to clearly separate what has been used for K_d derivation from any further supporting arguments and "conservatism".
  - It was stressed that the experts must be aware of the assumptions implicit in the K_d approach, such as reversibility. Within many experimental setups of hydrated cement systems the sorbing solid is undergoing alterations while sorption is proceeding but this apparent non-reversibility may have nothing to do with irreversible sorption in the thermodynamic sense. It was agreed that data for conditions other than those of interest are of limited use. However, it may be difficult to find K_d values directly relevant to the repository conditions. The challenge is then to extrapolate from the available data to repository conditions.
  - There were concerns raised over the use of thermodynamic data in linear free-energy models. However, it was suggested that some thermodynamic modelling would be necessary.

- *Appropriateness*: for the Dessel disposal facility: the review had to focus on relevant cement systems, or more generally relevant solid phases.
- *Data quality*: influence of solubility, appropriate S/L ratio, appropriate initial concentration of radionuclides, etc.
- *Time frames*: all four states of cement degradation were considered in the data selection because the cement, when in contact with infiltrating soil water and soil air, will experience each of these states within the lifetime of the repository; for the purpose of data selection knowledge of the start and end times of the degradation states is not necessary.
- *Heterogeneity*: several types of heterogeneity needed to be considered. These include:

  - Stable elements that need to be considered. The distribution of radioactive and non-radioactive wastes over the facility might be important, especially in view of isotopic dilution as a sorption process.
  - Chemical and/or microbiological degradation of bitumen might affect redox conditions, while reduction of nitrates leached from bituminised waste may further alter the redox state.
  - Organics within the waste packages and the potential impact of their degradation products on radionuclide migration.
  - Waste containing high chloride concentrations, on account of chlorocomplexation.
  - Water held in the interstitial pores in cement gel may not have the same thermodynamic properties as bulk water. Therefore, sorption values derived from bulk water experiments may not be applicable to gel pores.

- *Completeness*: for each species, possible mechanisms explaining the sorption processes were discussed, thereby seeking consistency among chemically similar species.

3 Selected Sorption Values

Scientific best estimate R_d values for each of the four concrete degradation states are given in Table 10.1 for 25 elements [10, 11]. These data represent sorption on unperturbed cement pastes. Sorption of radionuclides onto cement may be

Table 10.1 Selected best estimate R_d (ℓ/kg) values as function of cement degradation state

Element		State I	State II	State III	State IV
Ag		i.d. [1]	i.d. [1]	i.d. [1]	1
Am		1×10^4	1×10^4	1×10^4	1×10^4
Be		i.d.	i.d.	i.d.	i.d.
C[a]		2×10^3	5×10^3	2×10^3	i.d.
Ca		7	1	10	40
Cl		2×10^1 (1)	5×10^1 (1)	2×10^1 (1)	0 (0)
Cs		i.d.	2	20	i.d.
H		0	0	0	0
I		1 (1)	10 (1)	1 (1)	0.4 (0)
Mo		3	3	3[b], 0.1[c]	0.1
Nb		5×10^4	5×10^4	5×10^4	5×10^2
Ni		65	4×10^2	4×10^2	5
Np	Np (IV)	3×10^4	3×10^4	3×10^4	3×10^2
	Np (V)	i.d.	i.d.	i.d.	i.d.
Pa	Pa (IV)	3×10^4	3×10^4	3×10^4	3×10^4
	Pa (V)	1×10^4	1×10^4	1×10^4	3×10^2
Pb		3×10^2	3×10^3	3×10^4	8×10^2
Pd		i.d. [3×10^2]	i.d. [3×10^3]	i.d. [3×10^4]	i.d. [8×10^2]
Pu	Pu (IV)	5×10^3	3×10^4	3×10^4	3×10^2
	Pu (VI)	2×10^3	3×10^4	3×10^4	50
Ra		3×10^2	1×10^2	8×10^2	1
Se(−II)		i.d. [0]	i.d. [0]	i.d. [0]	i.d. [0]
Se(+IV)		2×10^2	2×10^2	2×10^2	3
Se(+VI)		3	3	3[b], 0.1[c]	0.1
Sn		2×10^4	2×10^4	i.d. [3]	i.d. [3]
Sr		1×10^2	30	1×10^2	1
Tc	Tc (IV)	3×10^3	3×10^3	3×10^3	i.d.
	Tc (VII)	1	1	1	1
Th		3×10^4	3×10^4	3×10^4	3×10^4
U	U (IV)	3×10^4	3×10^4	3×10^4	3×10^4
	U(VI)	2×10^3	3×10^4	3×10^4	50
Zr		i.d.	1×10^4	1×10^5	i.d.

Values in [] are supplemental values. For Ni and Ca site-specific cement data were used to calculate R_d for states I, II, and III. Values between parentheses are for high chloride background concentrations

i.d. insufficient data

[a]Inorganic carbon

[b]Ettringite present

[c]Ettringite absent

negatively affected by the presence of organic degradation products or high chloride concentrations. Corrections will need to be made for different concrete formulations including aggregates by accounting for the weight percent of cement.

4 Effect of Perturbing Components

Perturbing compounds exist in some waste streams or originate from cement and concrete as additives. These compounds include chloride and organics such as cellulose [with its degradation product, isosaccharinic acid (ISA)] from waste and cement admixture from concrete. The perturbing compounds may interact with radionuclides in the waste forming soluble complexes and as a result may decrease sorption on cement and increase the mobility of radionuclides. In general, the current review found very limited data enabling a quantitative assessment. A qualitative evaluation is, however, made and summarised hereafter:

- *Effect of organics*: for the majority of waste packages where ISA concentration is lower than 10^{-5} molal, no impact of organics on sorption is expected with the exception of Zr where a sorption reduction factor about 30 was observed. For individual waste packages with an ISA content between 10^{-4} and 10^{-3} molal, a sorption reduction factor of 10 was assigned to Nb, Np (IV), Pa, Th and a factor of 100 to Pb, Pu, and Zr. As concerns the high molecular weight organics, no effects have been reported based on experimental data for Pu. As concerns cement admixtures, small increase of Pu solubility was found.
- *Effect of chloride*: three groups can be distinguished based on the tendency of interaction with chloride or competing for sorption sites on cementitious materials
  - High impact: if chloride is present as NaCl, sorption reduction is expected for Cs, Sr, and Ra mainly due to Na^+ competition. Ag, Pb, and Pd belong to the high impact group also because of high tendency of forming complexes with chloride.
  - Medium impact: Tc(IV) as there is evidence for formation of Cl complexes, Ni (due to its position in the periodic table), Cl, I, Mo, and Se(VI) due to competing with Cl^-, and CO_3^{2-} for sorption sites on cement.
  - Low impact: U(VI), U(IV), Pu(VI), Th, Np(IV), Nb, Pa(V), Be, Ca, Se(IV), Sn, and Zr. These are not expected to interact strongly with chloride.

Acknowledgments This work has been performed as part of the project on disposal of category A waste—short-lived low and intermediate level waste (LILW-SL)—that is carried out by ONDRAF/NIRAS, the Belgian Agency for Radioactive Waste and enriched Fissile Materials.

Disclaimer The following members of the panel of experts are sincerely acknowledged for their contributions to the data review and endorsement: E. Giffaut, C. Gallé, V. Blin, B. Schwyn, A, Hooper, S. Williams, E. Wieland, D, Lever, and F. Glasser.The findings and conclusions in this paper are those of the authors and do not necessarily represent the official position of ONDRAF/ NIRAS. The work presented in this paper was more particularly carried out as preparatory work to a future license application file for a near surface disposal facility at Dessel to be submitted by

ONDRAF/NIRAS to the FANC. Therefore, the results in this paper have not yet undergone regulatory review nor approval by the FANC. Furthermore, it is not excluded that further developments and refinements of the work presented in this paper will be included in the future license application file.

References

1. ONDRAF/NIRAS 2010 Development of the safety concept and status mid-2010 of the disposal facility design, NIROND-TR 2007-03 V2 E, June 2010
2. Bradbury MH Sarott F (1995) Sorption databases for the cementitious near-filed of a L/ILW repository for performance assessment, PSI Bericht-95-06, 1995
3. Bradbury MH, Van Loon LR (1998) Cementitious near-field sorption data bases for performance assessment of a L/ILW disposal facility in a Palfris Mar Host Rock, CEM-94: Update I, June 1997, PSI Bericht, Nr. 98-01
4. Wieland E, Van Loon L (2002) Cementitious near-field sorption data base for performance assessment of an ILW repository in Opalinus Clay, PSI Bericht Nr. 03-06
5. Allard B (1985) Radionuclide sorption on concrete, NAGRA NTB 85-21. NAGRA, Baden, Switzerland
6. Baker S, Oliver P, McCrohon R (2002) Near-field batch sorption studies – 1992 to 1998, AEAT/ERRA-0345
7. Mallants D, Volckaert G, Labat S (2003) Parameters values used in the performance assessment of the disposal of low level radioactive waste at the nuclear zone Mol-Dessel, Volume 2: Annexes to the data collection forms for engineered barriers. SCK•CEN-R-3521, rev. 1, December, 2003, Mol, Belgium
8. Pointeau I, Landesman C, Coreau N, Moisan C, Reiller P (2004) Etude de la retention chimique des radionucleides Cs(I), Am(III), Zr(IV), Pu(IV), Nb(V), U(VI) ET Tc(IV) par les materiaux cimentaires degrades, RT DPC/SECR 03-037 indice A, CEA/DEN/DPC/SECR/L3MR DO 53, Juin 2004
9. Garcia-Gutierrez M, Alonso de los Rios U, Missana T, Mingarro M, Granizo N, Grivé M, Gaona X, Colàs E, Duro L, Bruno J (2007) Estudio bibliografico sobre sorcion y difusion de radionuclideidos en cementos, hormigones y productos de corrosion en presencia de cementos. Enresa technical report PT-02/07, 180 p
10. Wang L, Martens E, Jacques D, De Cannière P, Berry J, Mallants D (2009) Review of sorption values for the cementitious near field of a near surface radioactive waste disposal facility, NIROND-TR 2008-23 E, April 2009
11. Ochs M, Vielle-Petit L, Wang L, Mallants D, Leterme B (2011) Additional sorption parameters for the cementitious barriers of a near-surface repository, NIROND-TR 2010-06 E, March 2011
12. Jacques D (2008) Benchmarking of the cement model and detrimental chemical reactions including temperature dependent parameters, Project near surface disposal of category A waste at Dessel, NIROND-TR 2008-0X E
13. Jacques D, Wang L, Martens E, Mallants D (2010) Geochemical evolution of concrete during long-term interaction with soil water. Cement Concr Res 40:1306–1313. doi: 10.1016/j.cemconres.2010.02.008

Chapter 11
On the Determination of the Diffusion Coefficient of Ionic Species Through Porous Materials

H. Mercado, S. Lorente, and X. Bourbon

1 Introduction

Whichever may be its origin, radioactivity can be dangerous for people and their environment. Hence, radioactive material and waste involve a risk due to radionuclides that they contain. In the case of nuclear wastes storage and disposal, several protection barriers are set up in order to ensure the safety of people working or living near storage centers. Different types of confinement elements, often made from cementitious materials, are to be constructed so that optimal storage conditions are satisfied. Any attempt to confine efficiently a radioactive substance, within an enclosure made from a cement-based material, will certainly lead to the analysis of transfer properties of this kind of materials. Indeed, these properties play a main part in the suitability assessment of a determined storage solution. The diffusion coefficient of ionic species through porous materials is an essential input parameter that allows to model transfer phenomena and to predict material behavior in time. A study is presented in this chapter on the determination of the aforementioned diffusion coefficient through two different types of concrete, using two different methods, namely an electrochemical impedance spectroscopy (EIS) test and an electrokinetic one [1]. Special focus is made on impedance spectra analysis by means of two different electrical models.

H. Mercado • S. Lorente (✉)
UPS, INSA, LMDC (Laboratoire Matériaux et Durabilité des Constructions), Université de Toulouse, 135 Avenue de Rangueil, 31077 Toulouse Cedex 04, France
e-mail: sylvie.lorente@insa-toulouse.fr

X. Bourbon
ANDRA, 1/7 Rue Jean Monnet, Parc de la Croix Blanche, 92298 Chatenay-Malabry, France

F. Bart et al. (eds.), *Cement-Based Materials for Nuclear Waste Storage*,
DOI 10.1007/978-1-4614-3445-0_11,

Table 11.1 Concretes characteristics

	CEM-I concrete	CEM-V concrete
W/C	0.43	0.41
Apparent density (kg/m^3)	2,330	2,285
Water porosity	0.13	0.14

Table 11.2 Composition of the pore solutions and ionic strength I (mol/m^3)

	Na^+	K^+	OH^-	I
CEM-I concrete	31.5	122.8	154.3	154
CEM-V concrete	70.6	173.9	244.6	245

2 Materials

The concrete samples were based on CEM-I and CEM-V cements with water-to-cement ratio of 0.40 and 0.43, respectively. These materials are based on the ANDRA prescriptions and their composition is a reference composition. They were cast into cylindrical molds of 11 cm in diameter and 22 cm in height, before being demolded after 24 h. The samples were kept in humid chamber for at least 6 months for each type of concrete. At the end of the curing period, the cylinders were sawed into 3 cm thick pieces. The characteristics of the two concretes are presented in Table 11.1, together with their apparent density and water porosity (obtained from the AFPC-AFREM protocol [2]). The ionic composition of the pore solution was measured by atomic adsorption spectroscopy after extraction. Because sulfate and calcium concentrations were very small, Table 11.2 presents only the main ionic species concentrations and the ionic strength I of each solution. The latter was calculated from Eq. (11.1).

$$I = \frac{1}{2} \sum_i z_i^2 c_i \tag{11.1}$$

where i is the ionic species, z is the charge number, and c is the concentration.

3 Experiments

3.1 *Electrochemical Impedance Spectroscopy*

Impedance measurements were performed with an impedance analyzer (HP 4294A). The amplitude of the sinusoidal voltage was 200 mV and the frequency range was 40–(110×10^6) Hz. We designed our own electrodes made of a PVC envelope protecting a layer of copper and a thin mattress of stainless steel. The test

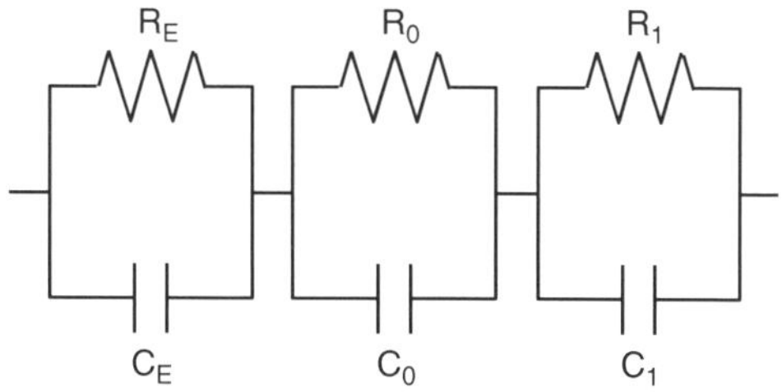

Fig. 11.1 Conventional electrical model

lasted <5 min per sample. The results are presented on Nyquist plots in terms of average of 3–4 tests, each test led in a day of interval, in a room at constant temperature.

The diffusion coefficient was obtained from the measurement of the saturated material resistance. Indeed the ratio of the pore solution conductivity σ_{pore} to the conductivity of the material saturated with the same pore solution σ_{mat} is the same as the ratio of an ion diffusion coefficient in the pore solution D_{pore} to the diffusion coefficient of the same ion though the saturated material, D. This ratio is known as the formation factor F. For more details on the application of the Nernst–Einstein relation and its applicability to cementitious materials, see for example Snyder's papers [3, 4].

$$F = \frac{\sigma_{pore}}{\sigma_{mat}} = \frac{D_{pore}}{D} \quad (11.2)$$

D_{pore} can be obtained in books and handbooks of chemistry, such as Atkins [5]. The material conductivity is calculated from

$$\sigma_{mat} = \frac{L}{AR_{mat}} \quad (11.3)$$

where L is the sample thickness (m), A is the sample cross-section (m^2), and R_{mat} is the saturated material resistance (Ω). The latter is measured from the EIS experiments.

3.1.1 Conventional Model

This model consists of representing the electrical characteristics of the material by two associations of a resistance and a capacitance placed in parallel. The electrodes are represented by a similar association. The three are assembled in series. The equivalent electrical circuit used in this work is shown in Fig. 11.1. The impedance of the equivalent circuit is given by:

$$Z(\omega) = Z_{real}(\omega) + i\, Z_{im}(\omega) \quad (11.4)$$

where

$$Z_{\text{real}}(\omega) = \frac{R_E}{1 + (\omega R_E C_E)^2} + \frac{R_0}{1 + (\omega R_0 C_0)^2} + \frac{R_1}{1 + (\omega R_1 C_1)^2} \tag{11.5}$$

and

$$Z_{\text{im}}(\omega) = \frac{\omega R_E^2 C_E}{1 + (\omega R_E C_E)^2} + \frac{\omega R_0^2 C_0}{1 + (\omega R_0 C_0)^2} + \frac{\omega R_1^2 C_1}{1 + (\omega R_1 C_1)^2} \tag{11.6}$$

The value of the AC angular frequency ω for which the imaginary part of the equivalent circuit impedance corresponds to a totally resistive behavior of the circuit is

$$Z(\omega) = R_{\text{mat}} \tag{11.7}$$

Yet, because the imaginary impedance is unlikely to reach a null value due to parasite effects, R_{mat} will be approached by the minimum value of Z_{im} [6]. Note that R_{mat} corresponds to R_0 + R_1.

3.1.2 Microstructure Model

If the preceding model is the most encountered in the literature, it presents the drawback of not being based on parameters with a physical meaning. Following the works published by Macphee et al. [7] or Song [8], we consider here that the porous material can be seen as a combination in parallel of three ways to transfer the electrical charges:

- The continuous conductive paths represent the fraction of opened and connected pores. They are represented by a resistance in the equivalent electrical model, R_{CCP}
- The solid matrix plays the role of an insulator, modeled by a capacitance C_{mat}
- The discontinuous pore paths are pore paths separated by cement paste layers

The association in series of a resistance R_{CP} and a capacitance C_{DP} represents this path where CP means continuous portion and DP discontinuous point. The microstructure model is then associated in series with the block representing the contribution of the electrodes as shown in Fig 11.2.

Writing that the two circuits of Figs. 11.1 and 11.2 are equivalent, in the sense that they must represent the same EIS experimental result, and omitting the electrode contribution we obtain

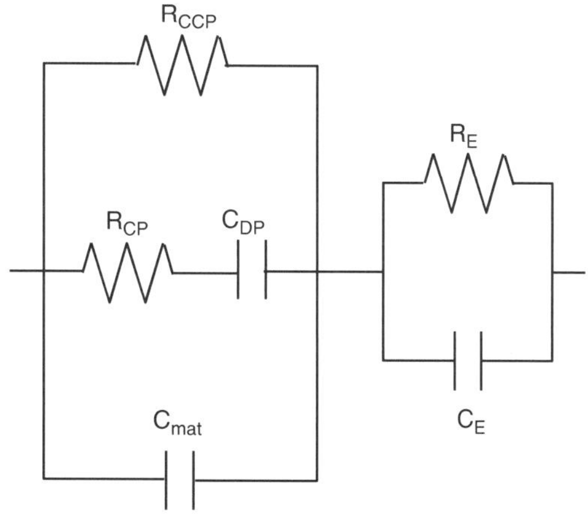

Fig. 11.2 Microstructure electrical model

$$\begin{aligned} C_{\mathrm{mat}} &= \frac{C_0 \cdot C_1}{C_0 + C_1} \\ C_{\mathrm{DP}} &= \frac{(R_0 \cdot C_0 - R_1 \cdot C_1)^2}{(R_0 + R_1)^2 (C_0 + C_1)} \\ R_{\mathrm{CCP}} &= R_0 + R_1 \\ R_{\mathrm{CP}} &= \frac{R_0 \cdot R_1 (C_0 + C_1)^2 (R_0 + R_1)}{(R_0 \cdot C_0 - R_1 \cdot C_1)^2} \end{aligned} \tag{11.8}$$

Experimentally, we note some dispersion on the results. In order to work on a more accurate modeling of the microstructure, the capacitances were replaced by constant phase elements (CPE) which are characterized by impedances such as

$$Z(\omega) = \frac{1}{Q(\omega i)^{\alpha}} \tag{11.9}$$

where Q is the value of 1/|Z| when $\omega = 1$ rad/s.

As a result the equivalent impedance of the entire circuit is

$$\begin{aligned} Z(\omega) = &\left(\frac{1}{R_{\mathrm{CCP}}} + \frac{1}{R_{\mathrm{CP}} + \frac{1}{Q_{\mathrm{DP}}(\omega i)^{\alpha_{\mathrm{DP}}}}} + Q_{\mathrm{mat}}(\omega i)^{\alpha_{\mathrm{mat}}} \right)^{-1} \\ &+ \left(\frac{1}{R_E} + Q_E(\omega i)^{\alpha_E} \right)^{-1} \end{aligned} \tag{11.10}$$

The optimization procedure consists of solving the equation

$$Z(\omega) - Z_{\mathrm{exp}}(\omega) = 0 \tag{11.11}$$

where the nine electrical characteristics (R, Q, α) are unknown. The optimization is based on the Simplex algorithm [9, 10]. The circuit component corresponding to the diffusion of species through the material is the continuous conductive resistance R_{CCP}.

3.2 Electrokinetic Tests

The principles of electrokinetic tests are well known and used in practice, in particular for the measurement of the chloride diffusion coefficient [11, 12]. The experimental setup used in this part of the work is a new version of the LMDC-Test (LMDC for Laboratory of Materials and Durability of Constructions) developed more than a decade ago in our laboratory.

A cylindrical sample of material is placed between two compartments containing any of the artificial solutions (corresponding to the pore solution analysis of each type of concrete) completed with 20 g/L of NaCl. Both compartments contain the same electrolytic solution. The cathodic side is filled with 500 mL of solution. The concrete samples were initially vacuum saturated with the same solution of NaOH, KOH, and NaCl. This means that no concentration gradient existed between the material and the electrolyte solutions. A constant electrical potential difference is applied between the two faces of the material in a way to create a DC 400 V/m electrical field. The geometry of the electrodes (made of stainless steel) ensures a uniform distribution of the electrical field through the sample, at least at the macroscopic level. The originality of the test lies in the measurement of the amount of chloride remaining in the cathodic compartment in time. The cathodic solution is regularly renewed in order to maintain constant boundary conditions. The test lasts <2 days and starts immediately after saturation, which should prevent from the changes in microstructure that can be noticed in time when classic migration tests are used [13]. The difference in the initial amount of chloride and the amount of chloride still in the cathodic compartment is the amount of chloride entering the material. The chloride concentration is measured by potentiometric titration. From these data, the flux of chloride can be calculated with Eq. (11.12).

$$J_i = -D_i z_i \frac{F}{RT} c_i \nabla\varphi \qquad (11.12)$$

where J is the flux of the ionic species i, D is its diffusion coefficient, c is the concentration in solution, z is the charge number, F is the Faraday's constant, R is the ideal gas constant, T is the absolute temperature, and φ is the electrical potential. Equation (11.12) is the Nernst–Planck equation written when all the concentration gradients are null, with therefore $\nabla\varphi = -\Delta U/L$, if we call ΔU the electrical potential difference and L the sample thickness [14, 15].

4 Results

We present in Fig. 11.3 the results obtained for the two types of concrete corresponding to the different cases of saturation: some samples were saturated with an artificial solution corresponding to the one extracted (Table 11.2). Other samples were saturated with the same artificial solution to which 20 g/L of NaCl (0.342 mol/L of Cl^-) were added. In this last case, the experiments were conducted after 3 weeks, 9 months, and 13 months during which the saturated samples remained immersed in their saturation solution.

Figure 11.3 shows both the experimental results and the theoretical ones (from the microstructure model). The optimization procedure results in a good agreement with the experiments. The low frequency range is dominated by the electrode behavior, while the results obtained in the highest frequencies are characteristics of the material itself. According to MacPhee et al. [7] the first loop would correspond to the pores system while the second one would give indications on the hydration products.

Figure 11.4 shows a zoom of Fig. 11.3b. The capacitance loop exhibits an inflexion which, in our view, corresponds to the superposition of two loops. This inflexion point is highlighted by the blue circle in the figure. According to some results in the literature [16], the impedance of the cementitious materials does not intersect the Nyquist diagram on its origin. The intersection is claimed to occur at a distance R from the origin. The corresponding equivalent model is therefore constituted of the resistance R placed in series with a "block" of resistance/capacitance in parallel. The physical explanation of such behavior varies from one paper to another, yet none of them appears to be really convincing. The frequencies used in this study are higher than those presented in the above cited papers. This may explain why, in our case, the measured impedances intersect the origin of the Nyquist diagram. This is also the reason why our experiments could also catch the second arc, measured between 5 and 110 MHz for both types of concrete.

The continuous conductive resistances R_{CCP}, result of the optimization procedure, are gathered in Table 11.3. Note that R_{CCP} is equivalent to R_{mat} in the conventional model.

Table 11.3 indicates that the continuous conductive resistance of a CEM-V concrete saturated with its own pore solution is twice as much as the R_{CCP} resistance of a CEM-I concrete saturated also with its own pore solution. This does not mean that the ratio between their diffusion coefficients is 2 since the diffusion coefficient as calculated with the Nernst–Einstein equation [Eq. (11.2)] depends also on the pore solution electrical properties.

Whatever the pore solution, the R_{CCP} resistance remains approximately constant in the case of CEM-I concrete which tends to show that even when chloride was present in the pore solution network for more than a year, it has very little impact on the results. The case of CEM-V concrete is different. The saturation in chloride of the pore solution leads to an increase of the continuous conductive resistance. This increase in R_{CCP} continues in time. Indeed the longer the presence of chloride in the pore solution the higher the resistance. This can be attributed to the change in

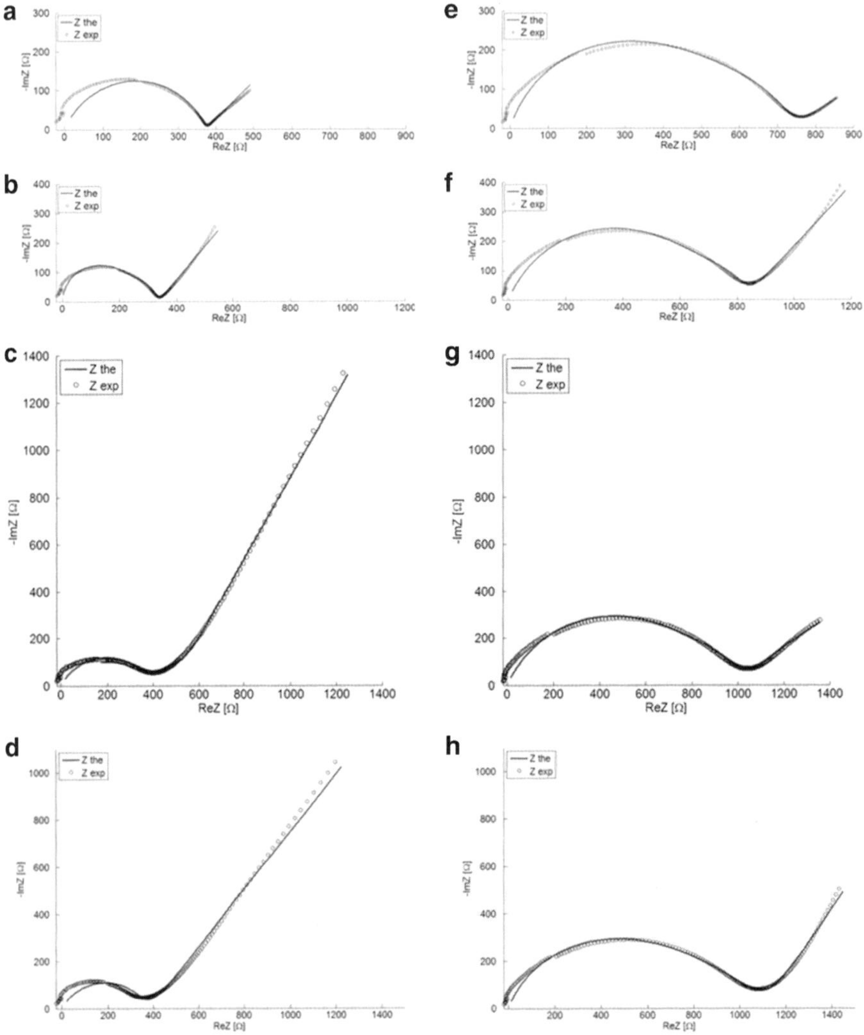

Fig. 11.3 EIS on concrete samples results. CEM-I concrete saturated with respectively its own pore solution (**a**), and NaCl for 3 weeks (**b**), and NaCl for 9 months (**c**), and NaCl for 13 months (**d**). CEM-V concrete saturated with respectively its own pore solution (**e**), and NaCl for 3 weeks (**f**), and NaCl for 9 months (**g**), and NaCl for 13 months (**h**). *Dotted* and *thick curves* stand for experimental and theoretical results, respectively

geometry of the pore network. The reaction of Cl^- with the hydration products of the solid matrix, and particularly with the aluminates leading to Friedel's salt, induces a narrowing of the paths measured through the increase of the resistance. The CEM-I seems to be not sensitive to this, at least for the duration of the experiments, probably because of its pore size repartition much less centered on the nanopores unlike CEM-V.

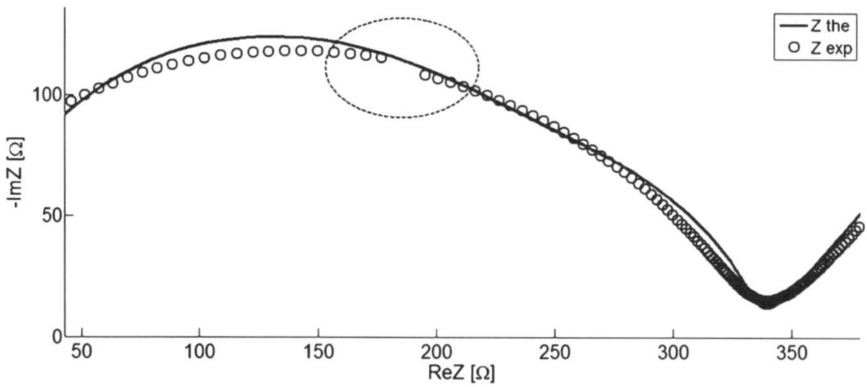

Fig. 11.4 Detail of Fig. 11.3, inflexion on curve (b)

Table 11.3 Continuous conductive resistances R_{CCP} (Ω)

	Pore solution	Pore solution + NaCl, 3 weeks	Pore solution + NaCl, 9 months	Pore solution + NaCl, 13 months
CEM-I concrete	373	334	496	387
CEM-V concrete	731	827	1,002	1,133

Table 11.4 Diffusion coefficients (m^2/s)

Saturation with...	The pore solution (EIS)	The pore solution + NaCl, 3 weeks (EIS)	The pore solution + NaCl, 3 weeks (Electrokinetics)
CEM-I concrete	4.1×10^{-12}	3.0×10^{-12}	2.8×10^{-12}
CEM-V concrete	1.5×10^{-12}	1.0×10^{-12}	0.8×10^{-12}

We present in Table 11.4 the diffusion coefficients corresponding to a saturation with NaOH–KOH, and to a saturation with NaOH–KOH and NaCl, the samples remaining in this last case in the saturation solution for 3 weeks.

A good agreement exists between results obtained with the EIS method and the electrokinetic one when the samples have been saturated with their respective pore solution and NaCl for 3 weeks. The diffusion coefficient of the CEM-I concrete remains relatively constant whatever the saturation solution, while the CEM-V concrete diffusion coefficient decreases in a more significant way after 3 weeks of saturation with chloride. This behavior is attributed to the partial blockage of the pore network by the reaction products of chloride with the cement matrix.

5 Concluding Remarks

We presented in this chapter an attempt to describe how the microstructure of two kinds of concrete has effect on the value of the diffusion coefficient, with an equivalent electrical model fitting EIS measurements. The microstructure model allows proposing a description of the results with physical meanings, unlike the conventional model. In particular, the continuous conductive resistance is the main indicator of the transports of ionic species. Its increase with the time of residence of chloride in the pore networks gives hints on the evolution of the microstructure in the case of CEM-V concrete and helps explaining the decrease in the chloride diffusion coefficient: features that the conventional electrical model cannot catch.

Finally, the comparison between this new technique of aqueous diffusion coefficient measurement and the electrokinetic (migration) technique is very good and makes EIS a good candidate for direct and fast measures.

Acknowledgments This work was supported by a contract from Agence Nationale pour la Gestion des Déchets Radioactifs (ANDRA).

References

1. Mercado H, Lorente S, Bourbon X (2011) Chloride diffusion coefficient: a comparison between impedance spectroscopy and electrokinetics test. Cement Concr Compos 34:68–75
2. AFPC-AFREM (1997) Méthodes recommandées pour la mesure des grandeurs associées à la durabilité, Toulouse
3. Snyder KA, Ferraris C, Martyrs NS, Garboczi EJ (2000) Using impedance spectroscopy to assess the viability of the rapid chloride test for determining concrete conductivity. J Res NIST 105:497–509
4. Snyder KA (2001) The relationship between the formation factor and the diffusion coefficient of porous materials saturated with concentrated electrolytes: theoretical and experimental considerations. Concr Sci Eng 3(12):216–224
5. Atkins PW (1998) Physical chemistry. Oxford University Press, Oxford
6. Barsoukov E., MacDonald JR (2005) Impedance spectroscopy, 2nd edn. John Wiley & Sons, Inc. Hoboken
7. Macphee DE, Sinclair DR, Cormack SL (1997) Development of an equivalent circuit model for cement pastes from microstructural considerations. J Am Ceram Soc 80:2876–2884
8. Song G (2000) Equivalent circuit model for AC electrochemical impedance spectroscopy of concrete. Cement Concr Res 30:1723–1730
9. Nelder J, Mead R (1965) A simplex method for function minimization. Comput J 7:308–313
10. Olsson D, Nelson L (1975) The Nelder-Mead simplex procedure for function minimization. Technometrics 17:45–51
11. AASHTO T 277-93 (1993) Standards specifications for transportation materials and methods of sampling and testing, 16th edn. Washington, DC, p 876
12. ASTMC 1202-94 (1994) American society for testing and materials. Philadelphia, PA, p 620
13. Sanchez I, Novoa XR, de Vera G, Climent MA (2008) Microstructural modifications in Portland cement concrete due to forced ionic migration tests. Study by impedance spectroscopy. Cement Concr Res 38:1015–1025

14. Lorente S, Voinitchi D, Bégué-Escaffit P, Bourbon X (2007) The single-valued diffusion coefficient for ionic diffusion through porous media. J Appl Phys 101(2):024907
15. Voinitchi D, Julien S, Lorente S (2008) The relation between electrokinetics and chloride transport through cement-based materials. Cement Concr Compos 30:157–166
16. Vedalakshmi R, Saraswathy V, Song Ha-Won, Palaniswamy N (2009) Determination of diffusion coefficient of chloride in concrete using Warburg diffusion coefficient. Corros Sci 51:1299–1307

Chapter 12
Moisture Transport Properties of Cement-Based Materials for Engineered Barriers in Radioactive Waste Disposal

X.Y. Pang, K.F. Li, and C.Q. Li

1 Introduction

Cement-based materials can be used for the engineered barriers for near surface disposal structures for radioactive wastes at low and intermediate levels [1]. These disposal structures are exposed to natural (underground) environments thus can be degraded, through physical processes or chemical reactions, by the external aggressive agents [2]. For cement-based materials, most of these degradation processes are intimately related to moisture (water) transport in the pore structure of materials [3]. Moisture, on one side, serves as the medium for aqueous transport of ions and other solutes, and, on the other side, can participate in the deterioration processes like leaching of the solid skeleton of materials. Thus mastering the moisture transport is fundamental for all other durability performance of cement-based engineered barriers. The moisture transport processes in concrete as a porous material have been investigated extensively so far in the literature [4–6]. In near surface disposal facilities, the concrete materials can be subject alternatively to water and dry conditions due to underground water level change and surface water infiltration. This paper introduces a multi-phase transport model of moisture in porous cement-based materials; with the measured transport properties the model is used to calculate the internal moisture distribution in cement-based engineered barriers exposed to natural drying and drying–wetting actions.

X.Y. Pang • K.F. Li (✉) • C.Q. Li
Key Laboratory of Safety and Durability of Civil Engineering of China
Education Ministry, Tsinghua University, Beijing, China

Civil Engineering Department, Tsinghua University, Beijing 100084, China
e-mail: likefei@tsinghua.edu.cn

F. Bart et al. (eds.), *Cement-Based Materials for Nuclear Waste Storage*,
DOI 10.1007/978-1-4614-3445-0_12,

2 Moisture Transport Modeling

2.1 Fundamental Properties for Moisture Transport

For moisture transport in a porous medium, the water storage characteristic is the fundamental property, i.e., the relationship between the external humidity h and the pore water saturation degree θ. The humidity h ($0 \leq h \leq 1$) is defined as the vapor pressure in pores divided by standard atmospheric pressure and the saturation degree θ ($0 \leq \theta \leq 1$) is defined as the volumetric ratio of liquid water in pores. For a porous medium, the saturation θ can be linked to the capillary pressure in pores, p_c, through the relation p_c–θ, also named as characteristic curve of moisture transport. For low permeable porous materials like concrete, a relation, adapted from the expression from Van Genuchten [7], is proposed by Mainguy et al. [8],

$$p_c(\theta) = a(\theta^{-b} - 1)^{1-\frac{1}{b}} \tag{12.1}$$

with a, b as experimental coefficients. Through perfect gas law and thermodynamic equilibrium at liquid–gas interface, the pressure difference across the vapor–water interface can be expressed as

$$\mathrm{d}p_l = \rho_l \frac{RT}{M_v} \mathrm{d}\ln(p_v) \tag{12.2}$$

where $p_{l,v}$ stand for liquid and vapor pressures (Pa), R is the perfect gas constant (J/mol/K), M_v is the molar mass of vapor (kg/mol), T is the temperature (K) and ρ_l is the liquid water density (kg/m^3). Taking vapor–liquid equilibrium of bulk water and saturated vapor under standard atmospheric pressure as reference state, the pore capillary p_c is expressed in terms of h through,

$$p_c(\theta) = a(\theta^{-b} - 1)^{1-\frac{1}{b}} = \rho_l \frac{RT}{M_v} \ln(h) \tag{12.3}$$

The relation θ–h from this expression is the moisture isotherm for a porous material.

2.2 Multi-phase Modeling of Moisture Transport

In multi-phase modeling of moisture transport, three phases are assumed to participate in pore transport: water (liquid), vapor (gas) and dry air (gas). This mixture transports in porous space of materials due to pressure (for liquid phase) or concentration gradient (for gas phases). It is also assumed that thermodynamic equilibrium is kept among the phases during the transport process. With these assumptions, the mass flux of vapor phase, J_v [kg/(m^{2}s)], writes [9]

$$J_v = -D_{va}\zeta\nabla\rho_v, \quad \nabla\rho_v = -\frac{M_v\rho_v}{\rho_l RT}\nabla p_c \tag{12.4}$$

In the above expression, D_{va} denotes the vapor diffusivity in air (m^2/s) and ζ the resistance of pore structure to gas diffusion (−) [10]. The second part of equation expresses the density in terms of vapor pressure through perfect gas state equation. The mass flux of liquid water, J_l [kg/(m^{2}s)], in porous medium is formulated by Darcy's law,

$$J_l = K_s K_r(\theta)\nabla p_c \tag{12.5}$$

where $K_{s,r}$ are saturate permeability (s) and relative permeability (−). The mass conservation of moisture in pores gives

$$\frac{\partial(\rho_l\varphi\theta)}{\partial t} = \nabla\cdot(J_v + J_l) \tag{12.6}$$

Finally the moisture transport is described by a diffusion-like equation

$$\frac{\partial\theta}{\partial t} = \nabla\cdot[D(\theta)\nabla\theta] \tag{12.7}$$

with the moisture diffusivity expressed as

$$D(\theta) = \left[K_s K_r(\theta) + D_{va}\zeta\frac{M_v\rho_v}{\rho_l RT}\right]\frac{\partial p_c}{\partial\theta} \tag{12.8}$$

The factor ζ assumes the expression $\zeta = \varphi^{2.7}(1 - \theta)^{10/3}$. And the relative permeability is expressed, based on the assumption of Van Genuchten [7], through

$$K_r(\theta) = \theta^{1/2}\left[1 - \left(1 - \theta^b\right)^{1/b}\right]^2 \tag{12.9}$$

So far, Eqs. (12.7)–(12.9) give a complete multi-phase modeling for moisture transport with boundary conditions prescribed on the water saturation.

2.3 Hysteresis Effect

From the above multi-phase model, one key point is the characteristic curve, i.e., p_c–θ relation, which determines the expression of moisture diffusivity in Eq. (12.8). However, experimental results show that this characteristic curve does not follow the same path for absorption and desorption processes [11]. This phenomenon is named as hysteresis of moisture transport. It is commonly accepted that the observed hysteresis is mainly due to the local connectivity of pores of different sizes, named as "ink-bottle" effect [12]. Actually, this non-uniqueness of p_c–θ relation has impact on the moisture diffusivity evaluated from Eq. (12.8),

$$D_{d,w}(\theta) = \left[K_s K_r(\theta) + D_{va}\zeta \frac{M_v \rho_v}{\rho_l RT}\right] \frac{\partial(p_c)_{d,w}}{\partial\theta} \tag{12.10}$$

In other words, the moisture diffusivity should be associated with different p_c–θ relations during drying (desorption) and wetting (adsorption).

3 Pore Structure and Sorption Isotherm

3.1 Material and Experiments

A C50 concrete, intended to be used as engineered barrier materials for near-surface waste disposal containers, is retained here for study. This concrete incorporates 45 % slag in the binder materials. The pore distribution of this concrete, measured on particles of hardened mortar, was evaluated through MIP test at age of 90 days, and the total porosity was evaluated form gravimetry measurement between saturated and totally dried states. Then the moisture isotherm was measured under six different controlled humidity environments under constant room temperature (20°C). The material mixture, as well as its measured properties, is given in Table 12.1.

3.2 Pore Structure and Sorption Isotherm

The accumulated pore volume and pore size distribution (PSD) from MIP are showed in Fig. 12.1, the measured moisture isotherm is illustrated in Fig. 12.2 and the characteristic curves are given in Fig. 12.3. The adsorption and desorption curves in Fig. 12.2 are respectively measured from totally dry and saturated samples

Table 12.1 Mix proportioning of concrete and the properties

Proportioning/property	Value
Cement PO42.5 (kg/m^3)	237
Slag (kg/m^3)	200
Coarse aggregates (5–20 mm) (kg/m^3)	1,055
Fine aggregates (<5 mm) (kg/m^3)	702
Water (kg/m^3)	175
Water to binder ratio w/b	0.40
Cubic strength at 28 days (MPa)	57.1
Capillary porosity by gravimetry	0.141

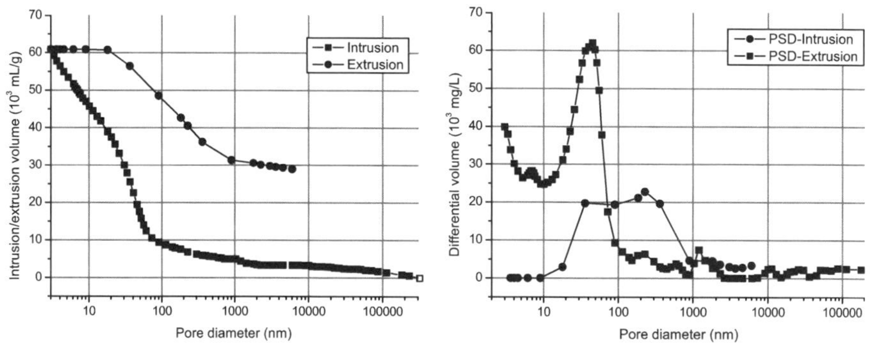

Fig. 12.1 Accumulated pore volume (*left*) and distribution (*right*) from MIP measurement

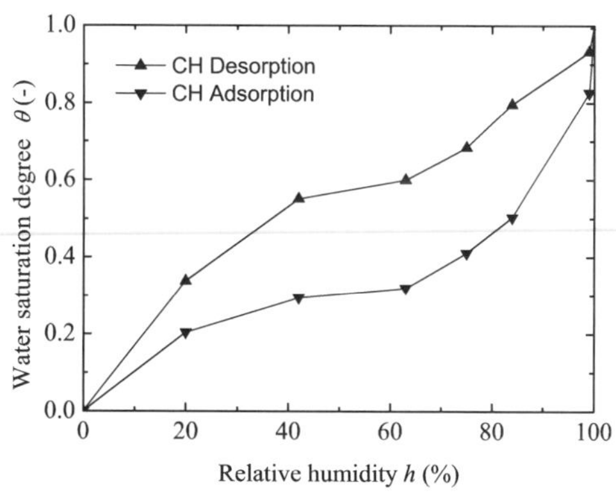

Fig. 12.2 Measured moisture isotherm for container concrete

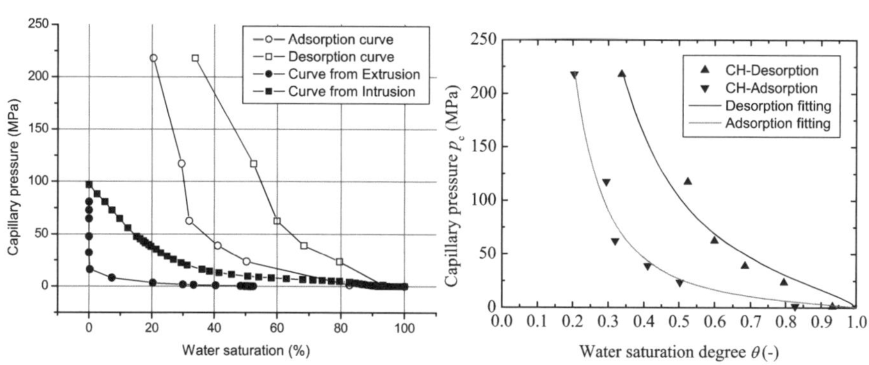

Fig. 12.3 Measured and predicted characteristic curves (*left*) and curve fitting results (*right*)

for six humidity conditions. From Fig. 12.1, one can see that the PSD is very different for mercury intrusion phase and extrusion phase and quite important quantity of mercury (~50 %) is retained in pores after extrusion. This reflects the fact that "ink-bottle" pores exist extensively in concrete and the MIP measurement gives just the size distribution of pore "neck" size.

Theoretically, the capillary pressure for moisture sorption can be deduced from the PSD,

$$p_c(\theta) = -\frac{2\gamma\cos\alpha}{r(\theta)}, \quad \theta = \frac{1}{V}\int_{r_{\min}}^{r} \frac{\mathrm{d}V}{\mathrm{d}\log r}\mathrm{d}r \tag{12.11}$$

where V denotes the total mercury intrusion volume, γ the surface tension of water in air (0.07275 N/m for 20°C) and α the contact angle between water and pore wall (180°). From this equation, the desorption and adsorption curves can be deduced respectively from the PSDs of the intrusion and extrusion phases.

The results are given in Fig. 12.3 (left) together with the measured curves. It can be seen that rather large difference exits between the predicted and the measured p_c–θ relations. The main cause is still the "ink-bottle" shape pores, which render the PSD from MIP not really the PSD. The cylindrical assumption for pores geometry in concrete is too simple to get realistic estimation for moisture transport properties. The second cause can be the MIP measurement on concrete. In fact, due to the sample size of MIP (several grams) and the heterogeneity of concrete materials on this scale, MIP cannot be reliable for concrete pore characterization. Accordingly, the measured isotherm is used in the multi-phase transport model and the curve fitting is presented in Fig. 12.3 (right). The regressed parameters are $a = 33.3$ MPa, $b = 2.77$ for desorption curve and $a = 6.02$ MPa, $b = 3.26$ for adsorption curve. The saturate permeability K_s is evaluated as 3.0×10^{-15} s and the diffusivity of vapor in air D_{va} is related as 2.45×10^{-5} m^2/s.

4 Moisture Transport in Cement-Based Engineered Barriers

In this section, the moisture transport is calculated for some typical environmental exposure conditions for disposal containers. The retained concrete is adopted as the structural material for the container in near-surface disposal.

The following cases investigated (1) a simple drying process with the initial water content as 0.8 and surface water content as 0.4 (corresponding to an external humidity of 30 % according to Fig. 12.2); (2) annual drying–wetting action with wetting period of 30 days; (3) annual drying–wetting action with wetting period of 60 days. The two latter cases simulate the annual contact of container wall with underground water movement. The calculation is effectuated for a container wall of thickness of 50 cm. The detailed boundary conditions are given in Table 12.2. The involved model is the multi-phase transport model in Sect. 2 and the involved parameters were presented in Sect. 3.

Table 12.2 Boundary conditions for three simulation cases

		Drying		Wetting		
Study case	Initial state	Period	Condition	Period	Condition	Cycle
1	$\theta = 0.8$	30 years	$\theta = 0.4$	–	–	–
2	$\theta = 0.8$	330 days	$\theta = 0.4$	30 days	$\theta = 1.0$	30
3	$\theta = 0.8$	300 days	$\theta = 0.4$	60 days	$\theta = 1.0$	30

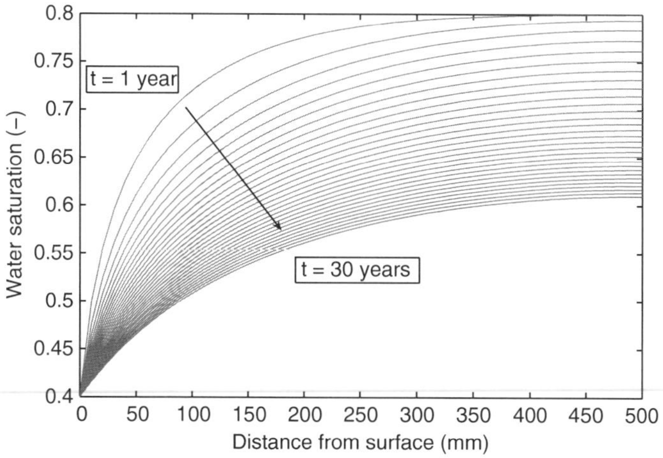

Fig. 12.4 Moisture profiles during drying for 30 years

The moisture profiles during simple drying of 30 years are presented in Fig. 12.4. It can be seen that the drying front has largely passed the wall thickness and final water saturation drops from 0.8 to about 0.6. The moisture profiles for wetting periods of 30 and 60 days are illustrated respectively in Figs. 12.5 and 12.6. From the two figures, it can be seen that under the drying and wetting actions the concrete surface alternatively losses water by drying and absorbs water by wetting and the final moisture profiles are determined by the two processes. A convection zone is identified on the two figures in which the water saturation changes actively with the boundary conditions. This convection zone can be very important for ion transport and the electrolytic process of steel corrosion. From the two drying–wetting cases, it can be seen that wetting period of 60 days gives a larger convection zone (about 180 mm) compared to the case of wetting period of 30 days (about 110 mm). Due to the longer drying period, the wall loses more water for 30 days wetting case than for 60 days wetting case.

Certainly these are only numerical results, and more calculations are needed to have a clear dependence law of convection depth on the initial water saturation, surface drying condition as well as the drying–wetting time ratio. A theoretical analysis of the convection zone can be found in [13].

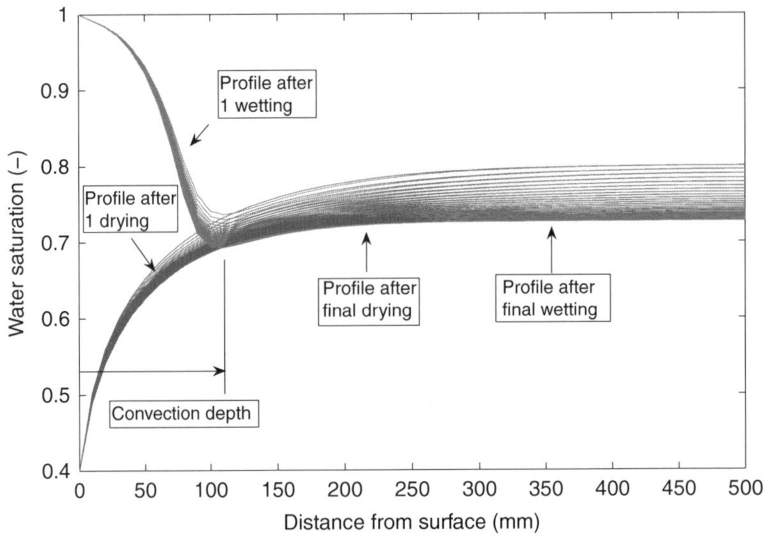

Fig. 12.5 Moisture profiles during drying–wetting actions of 30 years (wetting period 30 days)

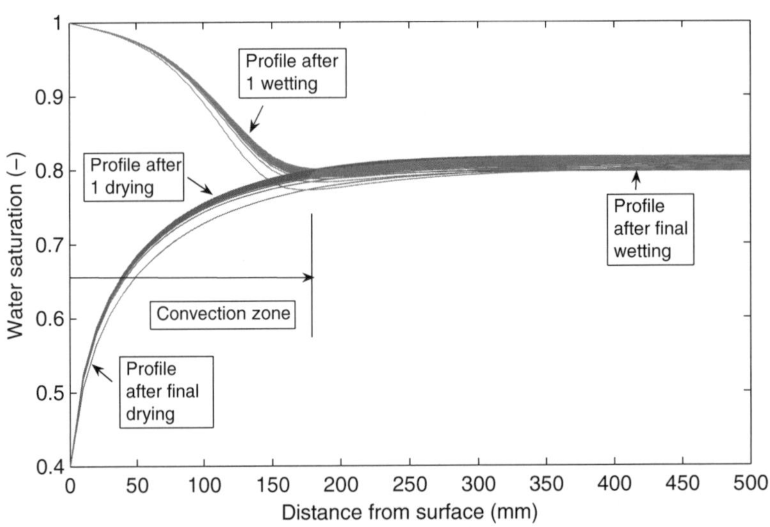

Fig. 12.6 Moisture profiles during drying–wetting actions of 30 years (wetting period 60 days)

5 Conclusions

1. From this research, the multi-phase modeling of moisture transport in concrete as partially saturated porous media is established for engineered barrier materials in radioactive waste disposal. The model considers the pore phases as liquid water, vapor and dry air phases and gives a diffusion-likc cquation for moisture transport using water saturation as basic variable. The key property of moisture transport is identified as the moisture isotherm or the characteristic curve.
2. A typical structural concrete for disposal container was investigated for its pore structure as well as its moisture transport properties. The pore structure measured by MIP shows a large difference between intrusion and extrusion phases and the retained mercury in pores is about 50 % of total intrusion volume, confirming the extensive existence of "ink-bottle" pores in material. Also the moisture isotherm predicted from PSD by MIP is far from the measured one.
3. Using the measured moisture transport properties, the concrete wall in disposal containers is exposed to different environmental actions including simple drying and drying–wetting alternative cycles. From the results, it can be seen that the drying fronts pass the wall thickness at the end of 30 years and a convection depth is observed for concrete surface exposed to drying and wetting actions. This convection depth depends strongly on the ratio between drying and wetting periods.

Acknowledgments This research is supported by China National Natural Science Foundation Grant No. 50978144 and China national major fundamental research grant (973 Program, No. 2009CB623106).

References

1. IAEA (2001) Performance of engineered barrier materials in near surface disposal facilities. IAEA-TECDOC-1225
2. IAEA (1999) Safety requirements: near surface disposal of radioactive waste. IAEA Safety Standards Series No. WS-R-1
3. Hobbs DW, Matthews JD, Marsh BK (1998) Minimum requirements of durable concrete: carbonation- and chloride-induced corrosion, freeze-thaw attack and chemical attack. British Cement Association, Crowthorne
4. Cerny R, Rovnanikova P (2002) Transport process in concrete. Taylor & Francis, London
5. Janssen H, Blocken B, Carmeliet J (2007) Conservative modeling of the moisture and heat transfer in building components under atmospheric excitation. Int J Heat Mass Transf 50:1128–1140
6. Thiery M, Baroghel-Bouny V, Bourneton N, Villain G, Stefani C (2007) Modélisation du séchage des bétons: analyse des différents modes de transfert hydrique. Revue Européenne de Génie Civil 11:541–578
7. Van Genuchten MT (1980) A closed-form equation for predicting the hydraulic conductivity of unsaturated soils. Soil Sci Soc Am J 44:892–898
8. Mainguy M, Coussy O, Baroghel-Bouny V (2001) The role of air pressure in the drying of weakly permeable materials. ASCE J Eng Mech 127:582–592

9. Philip JR, Vries DA (1957) Moisture movement in porous materials under temperature gradients. Trans Am Geophys Union 38:222–232
10. Millington RJ (1959) Gas diffusion in porous media. Science 130:100–102
11. Dullien FAL (1992) Porous media: fluid transport and pore structure. Academic, San Diego, CA
12. Espinosa RM, Franke L (2006) Ink bottle pore-method: prediction of hygroscopic water content in hardened cement paste at variable climatic conditions. Cement Concr Res 36:1954–1968
13. Li KF, Li CQ, Chen ZY (2009) Influential depth of moisture transport in concrete subject to drying-wetting cycles. Cement Concr Compos 31:693–698

Chapter 13
Simulation of Gas Hydrogen Diffusion Through Partially Water Saturated Monomodal Materials

C. Boher, S. Lorente, F. Frizon, and F. Bart

1 Introduction

Today, the preferred solution for low and intermediate level nuclear wastes consists of using cementitious materials as an envelope and physical barrier of the waste. In some cases, hydrogen may be produced as a result of radiolyse reactions. The risk is that hydrogen may accumulate to the limit of becoming explosive. It is therefore of great importance to design concrete envelopes which pore network allows the diffusion of hydrogen towards the outside. This work is part of a systematic study, both experimental and numerical, of H_2 diffusion through complex pore networks at various saturation degrees [1–3]. More detailed studies are also conducted on one-mode materials, i.e., materials with an almost unique peak in the pore size repartition.

C. Boher
UPS, INSA, LMDC (Laboratoire Matériaux et Durabilité des Constructions),
Université de Toulouse, 135 Avenue de Rangueil, 31077 Toulouse Cedex 04, France

Commissariat à l'Energie Atomique et aux Energies Alternatives, DEN/MAR/DTCD/SPDE,
Bagnols-sur-Cèze, France

S. Lorente (✉)
UPS, INSA, LMDC (Laboratoire Matériaux et Durabilité des Constructions),
Université de Toulouse, 135 Avenue de Rangueil, 31077 Toulouse Cedex 04, France
e-mail: sylvie.lorente@insa-toulouse.fr

F. Frizon
Commissariat à l'Energie Atomique et aux Energies Alternatives, DEN/MAR/DTCD/SPDE,
Bagnols-sur-Cèze, France

Laboratoire d'Etude de l'Enrobage des Déchets, French Atomic Energy and Alternative Energies Commission, BP17171, 30207 Bagnols sur Cèze Cedex, France

F. Bart
Commissariat à l'Energie Atomique et aux Energies Alternatives, DEN/MAR/DTCD/SPDE,
Bagnols-sur-Cèze, France

F. Bart et al. (eds.), *Cement-Based Materials for Nuclear Waste Storage*,
DOI 10.1007/978-1-4614-3445-0_13,

Table 13.1 Composition of the geopolymers (molar ratio)

		NaOH 0.6	NaOH 1.1	NaOH 1.2	NaOH 1.3
Global ratio	Na_2O/AlO_3	1	1	1	1
	SiO_2/AlO_3	3.0	3.5	3.6	3.7
	Na_2O/SiO_2	0.33	0.29	0.28	0.27
High pH solution	SiO_2/Na_2O	0.6	1.1	1.2	1.3

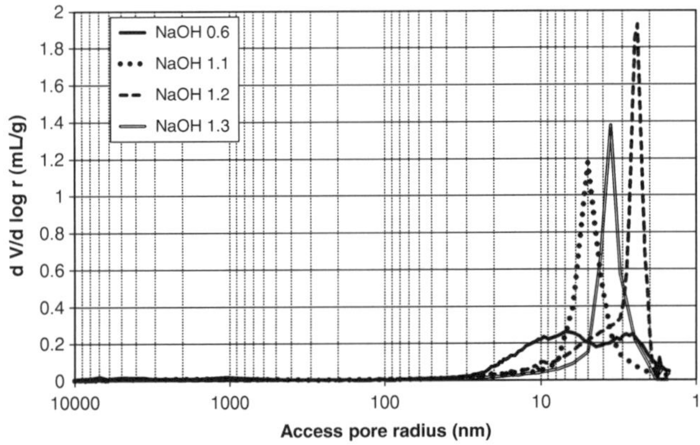

Fig. 13.1 Geopolymers pore size repartition (from mercury porosimetry)

The objective of this specific study is to determine the contribution of a pore size in the diffusion process. This chapter documents the modeling of a quasi one-mode material: geopolymers. According to their formulation, these materials can have a specific pore size repartition and are quasi monomodal. The study quantifies the pore network impact on the diffusion coefficient. The effect of the saturation level on the hydrogen diffusion coefficient is also highlighted.

The word "geopolymer" was introduced in 1978 [4] to describe aluminosilicate inorganics polymers. These materials are obtained from an aluminosilicate source, activated by an alkaline solution with a high pH.

The geopolymers studied here are made of metakaolin and amorphous silica as aluminosilicate source to which a solution of NaOH is added, as high pH source. The detailed compositions are given in Table 13.1.

Because of the polymerization reactions, these formulations have a very specific pore network. Figure 13.1 shows the pore size distribution, obtained with mercury intrusion porosimetry.

Depending on the formulation, the pore size range varies between 1.5 and 10 nm. This is very interesting because it allows us to model the hydrogen diffusion through a porous material with controlled pore size.

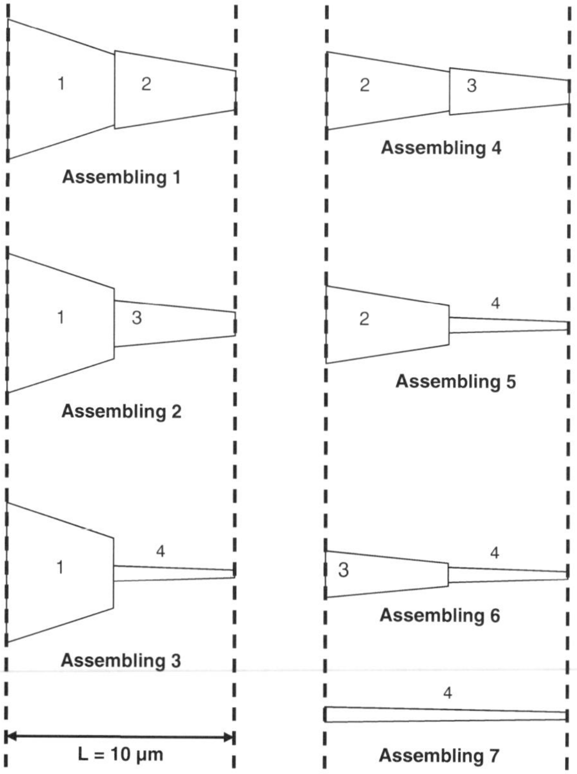

Fig. 13.2 Repartition of the pore families in the different assemblings

2 MOHYCAN Description

MOHYCAN is a model developed in our research group [5] which calculates the hydrogen diffusion coefficient through a virtual network, for saturation degrees ranging from 0 to 100 %.

MOHYCAN creates a virtual pore network from mercury porosimetry data. Since the previous version of the work, the accuracy of MOHYCAN virtual pore network was improved by discretizing into four pore families the pore network repartition obtained from mercury porosimetry. The four pores families have a length of L/2 and are combined into "Assemblings" that are depicted in Fig. 13.2.

The family number of the different pore families is assigned to have the following basic property:

$$N_{\text{pores,family } n^\circ 1} < N_{\text{pores,family } n^\circ 2} < N_{\text{pores,family } n^\circ 3} < N_{\text{pores,family } n^\circ 4} \tag{13.1}$$

where $N_{\text{pores, family } n^\circ x}$ is the number of the pore family $n^\circ x$.

Table 13.2 Presentation of the different cases of assembling proportions investigated

Case	A	B	C	D	E	F	G
P_1 (%)	0	0	10	25	50	75	100
P_2 (%)	0	10	10	25	50	25	100
P_4 (%)	0	10	10	25	50	100	100

From this initial assumption, each amount of Assembling is calculated according to the following algorithm:

- All the pores from family 1 are distributed between Assemblings 1, 2, and 3.
- The model user specifies the relative proportion of Assembling 1 and Assembling 2. Assembling 3 relative proportion is automatically calculated.
- The remaining pores from family 2 which have not been used in Assembling 1 are distributed between Assemblings 4 and 5.
- The user specifies the relative proportion of Assembling 4. Assembling 5 is automatically calculated.
- The remaining pores from family 3 which have not been used in Assembling 2 and Assembling 4 are distributed in Assembling 6.
- The remaining pores from family 4 which have not been used in Assemblings 3, 5, and 6 constituted the Assembling 7. The latter is made of the pores from family 4.

According to the previous algorithm, the model user must specify three proportions, noted:

- P_1 (proportion of Assembling 1)
- P_2 (proportion of Assembling 2)
- P_4 (proportion of Assembling 4)

In our study, seven proportions cases are investigated. They are described in Table 13.2:

- In case A, Assemblings 3, 5, 6 are privileged, i.e., Assemblings with the bigger number of pore family.
- In case G, Assemblings 1, 4, 6, 7 are privileged, i.e., Assemblings with the smallest number of pore family.
- Cases B–F make the transition between case A and case G.

When the virtual porous network is created, a mercury intrusion porosimetry is simulated in order to compare the artificial pore network to the experimental one. If the virtual pore network fits the experimental one, the model continues the calculation. Next, water layers with a constant thickness are deposited into the entire virtual network as a function of the expected saturation degree. The total water layer thickness varies from 0 nm to the maximal porous network radius. Thus, the water saturation level starts from 0 % and ends at 100 %.

For each saturation degree, the hydrogen diffusion coefficient through the artificial pore network is calculated, as described in the next paragraph.

Depending on the pore size, the diffusion mechanisms are different. Free molecules diffusion is dominant when the mean free path is negligible in front of the pore radius. On the other hand, when the pore radius is negligible in front of mean free path, diffusion is due to Knudsen regime. When the mean free path and the pore size have the same magnitude, the two regimes occur simultaneously in the diffusion process. To cover both diffusion mechanisms, the following equation is defined:

$$\frac{1}{D_{c,i}} = \frac{1}{D_{m,i}} + \frac{1}{D_{k,i}} \tag{13.2}$$

where $D_{c,i}$ is the global diffusion coefficient ($m^2\,s^{-1}$), $D_{m,i}$ and $D_{k,i}$ are, respectively, the free molecule diffusion coefficient and the Knudsen diffusion coefficient ($m^2\,s^{-1}$). Both of $D_{m,i}$ and $D_{k,i}$ are explained below [6].

$$D_{m,i} = \frac{3\pi}{16}\lambda_e\sqrt{\frac{kT}{m}}\left(\frac{kT}{\sqrt{2}\pi\sigma^2 P}\right) \tag{13.3}$$

$$D_{k,i} = \frac{2\lambda_e}{3}\sqrt{\frac{kT}{m}} \tag{13.4}$$

$$\frac{1}{\lambda_e} = \frac{1}{\lambda} + \frac{1}{r} \tag{13.5}$$

where λ_e (m) is the equivalent mean free path [7], k is the Boltzmann constant, $k = 1.3807 \times 10^{-23}\,JK^{-1}\,mol^{-1}$, T is the temperature (K), m is the mass (kg), σ is the molecule diameter (m), λ is the mean free path (m), and r is the pore radius (m).

3 Application to Geopolymers

Before running MOHYCAN in order to calculate the diffusion coefficient, mercury porosimetry data need to be discretized into four families, which boundaries are defined as:

- Rb1 is the largest pore radius with a $(dV/d(\log r)>0$.
- Rb2 is located at the abscissa where $(dV/d(\log r)) = 5\%$ of curve peak maxima.
- Rb3 and Rb4 are placed on the both sides of the peak, where $(d(dV/d(\log r)))/dr$ is maximum, i.e., where the peak slope is maximum.
- Rb5 is the smallest pore radius investigated with mercury porosimetry (=1.5 nm) (Fig. 13.3).

Once the boundary radii are chosen, the model can define the pore families and the diffusion coefficient is calculated as a function of the saturation degree. Figure 13.4 illustrates the virtual porous network results.

As shown in Fig. 13.4, there is a good correspondence between the experimental results and the modeled pore network. The hydrogen diffusion coefficient can be calculated through these networks for the various saturation levels. Figure 13.5 illustrates the results obtained as a function of the water saturation degree.

Figure 13.5 shows that the diffusion coefficient decreases in a quasi linear way when the saturation level increases until a threshold after which a sudden drop in the diffusion coefficient values is noticed.

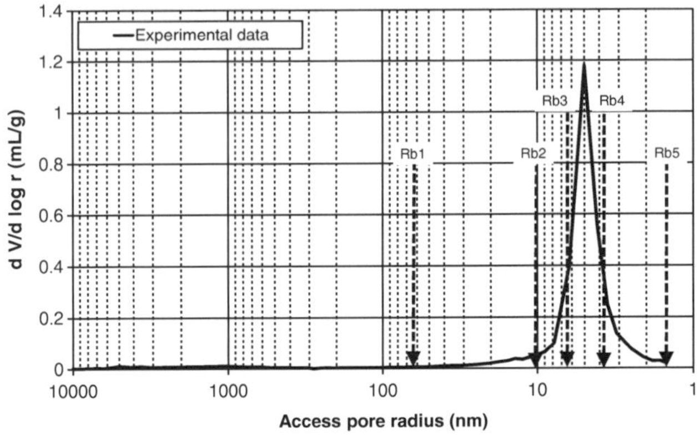

Fig. 13.3 Discretized pore size repartition, extracted from mercury porosimetry. Geopolymer with a formulation of NaOH 1.1

The simulations of the geopolymer NaOH 0.6 behavior indicate a discrepancy in the results when the saturation degree varies from 40 to 90 %. We see here the impact of the pore assemblings on the hydrogen diffusion coefficient evolution. The explanation of this result is to be found in the pore size distribution which is wider than the other geopolymers. This observation highlights the importance of the pore families assembling when the materials studied are not monomodal.

In the case where the porous materials are monomodal (NaOH 1.1, 1.2, and 1.3), the diffusion coefficient drop is more sudden than with other material (NaOH 0.6 for example). This is due to the complete water saturation of the main pore family. Therefore, because the hydrogen diffusion in water is 1,000 times slower than through gaz [8,9], we obtain a diffusion coefficient which is 6 orders of magnitude smaller.

4 Influence of Pore Family Number

The aim of this part is to study the impact of the mercury porosimetry discretization on the hydrogen diffusion coefficient. In this part, only the NaOH 1.2 geopolymer formulation will be studied.

Figure 13.5 details the discretization of the pore size repartition, depending on the pore family number (Fig. 13.6).

Once the experimental data are discretized, the virtual networks are created and the diffusion coefficients are calculated. The results are shown in Figs. 13.7 and 13.8.

According to Fig. 13.7, the more numerous the pore families, the better the match between experimental and modeled pore network. Using a one-pore-family discretization, the difference between the numerical and the experimental pore network is too important. Consequently, a one-pore family-based model cannot be used, because of its nonrealistic virtual pore network. This is due to the geopolymers pore network itself. Geopolymers are quasi-monomodal but not

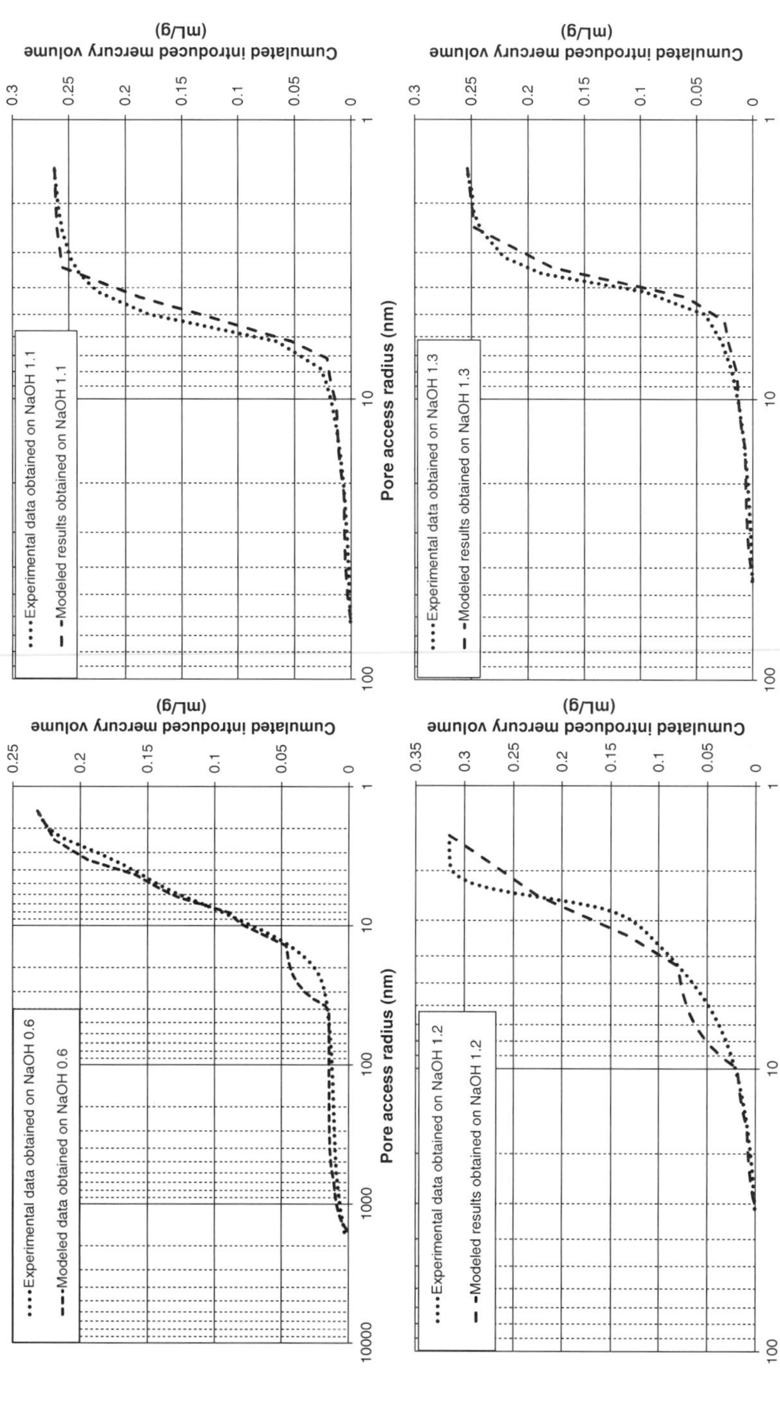

Fig. 13.4 Comparison of experimental and modeled cumulated introduced mercury volumes. Data obtained with geopolymers from NaOH 0.6, 1.1, 1.2, and 1.3

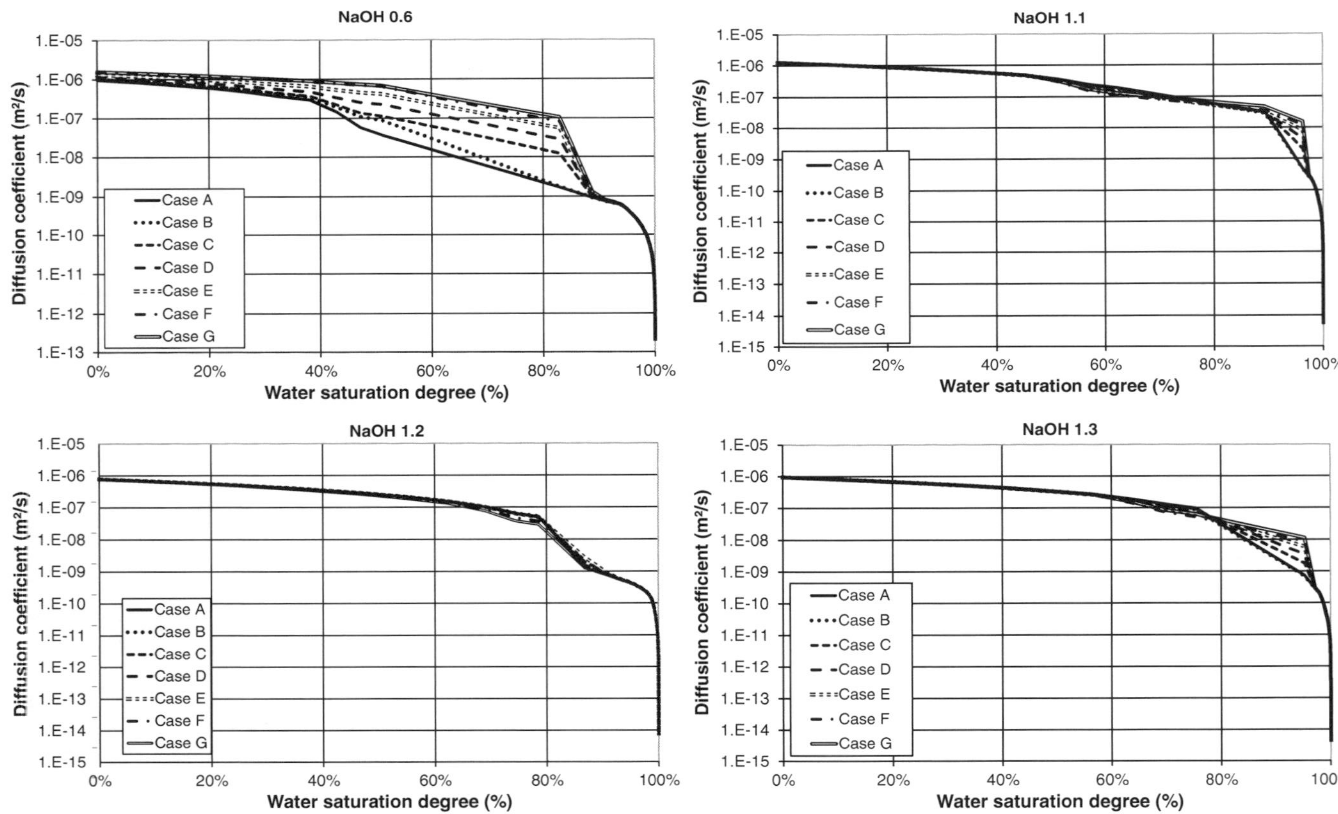

Fig. 13.5 Calculated diffusion coefficient as a function of the water saturation level. Results obtained on NaOH 0.6, 1.1, 1.2, and 1.3

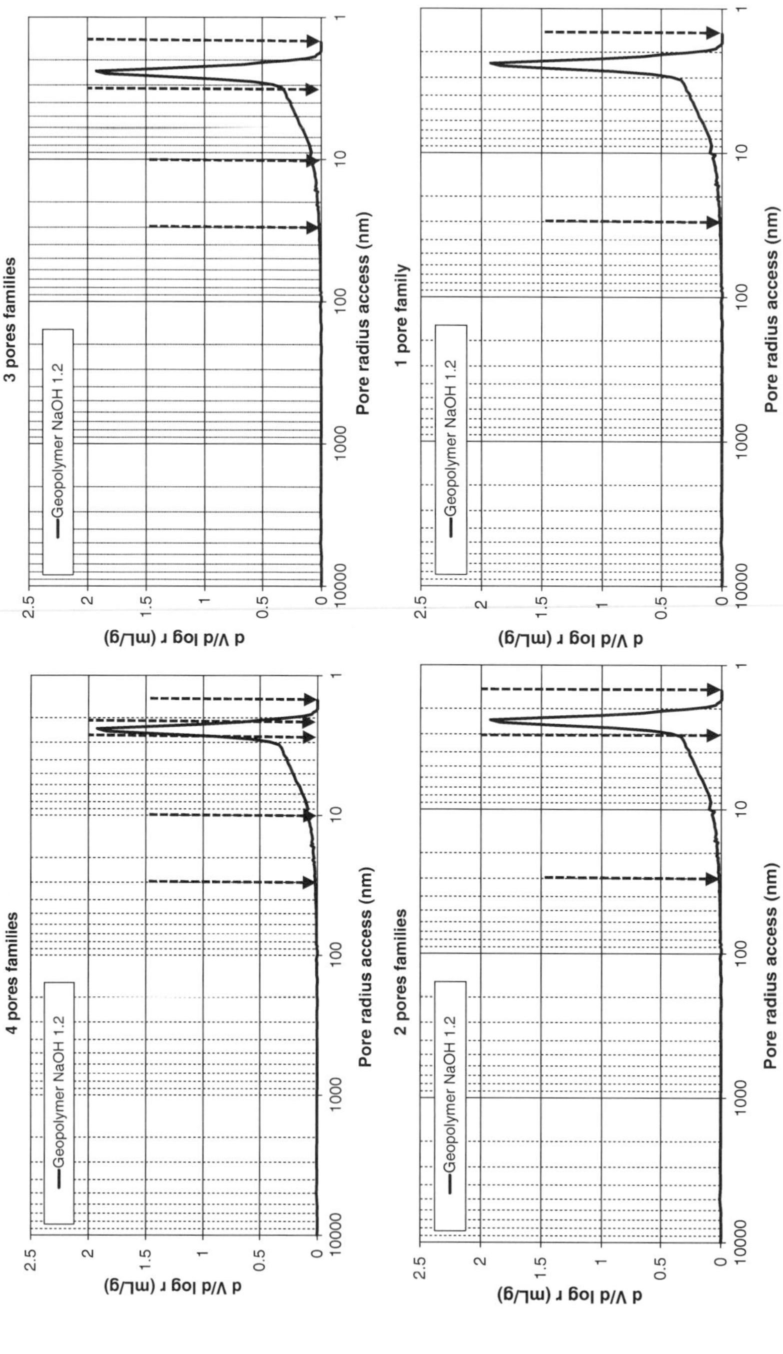

Fig. 13.6 Presentation of boundaries radii in the different case of pore families number (boundaries placed on pore size repartition obtained with mercury porosimetry)

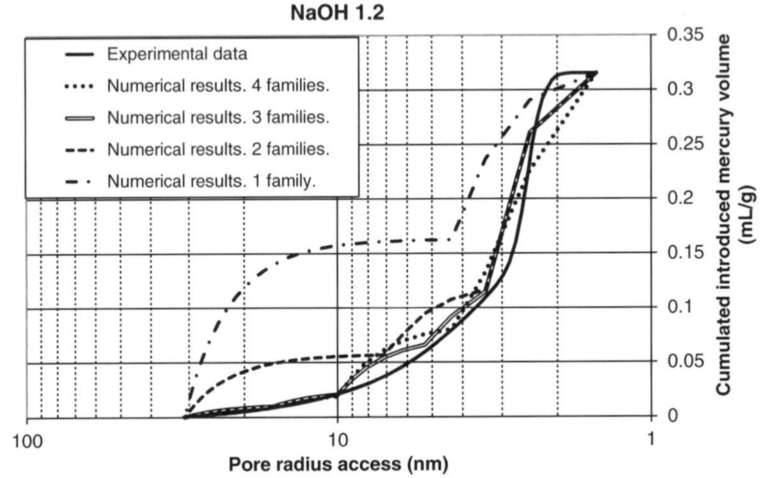

Fig. 13.7 Comparison between experimental and numerical pore size distribution

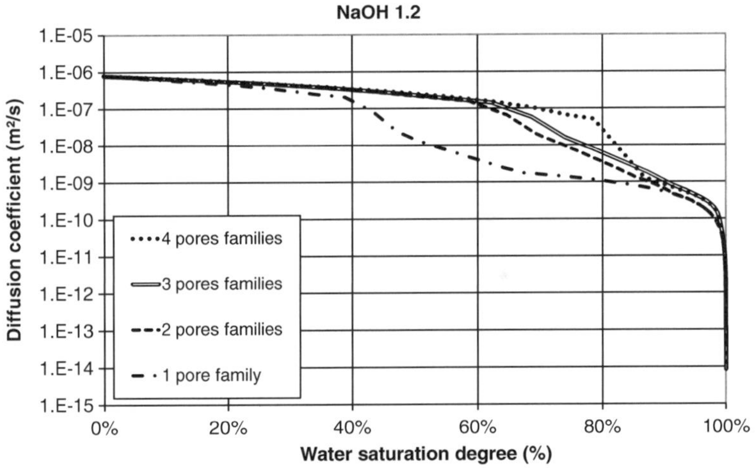

Fig. 13.8 Hydrogen diffusion coefficient as a function of water saturation degrees

perfect monomodal materials. The residual porosity on both sides of the peak a one-pore-family modelization.

The information given by Fig. 13.7 is confirmed by Fig. 13.8: diffusion coefficient calculated with a one-pore family network is away from the three others.

Concerning the two-, three-, and four-pore family-based models, results of simulated mercury porosimetry are acceptable approximation of the experimental one. Moreover, calculated diffusion coefficients are similar. So, in the case of monomodal materials, a three-pore family-based model is good enough to represent the real pore network.

5 Conclusion

In this chapter, we presented MOHYCAN, a model that calculates the hydrogen diffusion coefficient as a function of the water saturation degree, from mercury porosimetry data. This model is applied to a new material: the geopolymer. These materials are made from mineral polymerization in a high pH solution and are quasi monomodal.

After verifying that the modeled pore network reflects the actual one, diffusion coefficients are calculated according to the water saturation degree. The results are for a quasi monomodal material:

- The diffusion coefficient is not sensitive to different virtual pore network arrangement.
- The diffusion coefficient values have a sharp drop at specific water saturation. This is due to the water saturation of the main and unique pore family.
- A two-pore family-based model is sufficient to represent the pore network.

These observations will not be valid if we consider a material with a large pore size repartition, like cementitious materials. Yet we hope to be able in the next future to propose a good representation of this complex network by combining simple pore arrangements such as the ones encountered with geopolymers.

References

1. Sercombe J, Vidal R, Galle C, Adenot F (2007) Experimental study of gas diffusion in cement paste. Cement Concr Res 37:579–588
2. Frizon E, Galle C (2009) Experimental investigations of diffusive and convective transport of inert gas through cement pastes. J Porous Media 12:221–237
3. Vu TH (2009) Caractérisation de la phase solide et transferts de gaz dans les milieux poreux insaturés. Etude expérimentale et modélisation appliquées à la diffusion de l'hydrogène dans les matériaux cimentaires. INSA de Toulouse. Université de Toulouse, Toulouse
4. Davidovits J (1989) Geopolymers and geopolymeric materials. J Therm Anal 35:429–441
5. Vu TH, Frizon F, Lorente S (2009) Architecture for gas transport through cementitious materials. J Phys D Appl Phys 42:9
6. Aryanpour G, Abbasi MH (2005) Computer simulation of ordinary gas transfer in tubes. J Porous Media 8:379–391
7. Schüth F, Sing KSW, Weitkamp J (2002) Handbook of porous solids. Wiley, Weinheim
8. Cussler EL (1997) Diffusion. Mass transfer in fluid systems. Cambridge University Press, Cambridge
9. Bourgeat A, Jurak M, Smai F (2009) Two-phase, partially miscible flow and transport modeling in porous media; application to gas migration in a nuclear waste repository. Comput Geosci 13:29–42
10. Shaaf SA, Chamber PL (1961) Flow of rarefied gases. Princeton University Press, Princeton

Chapter 14
Microbial Catalysis of Redox Reactions in Concrete Cells of Nuclear Waste Repositories: A Review and Introduction

A. Albrecht, A. Bertron, and M. Libert

1 Redox Reactions in Natural Biogeochemical as Well as Anthropogenic Systems

Reduction–oxidation reactions are the corner stone of all life on Earth, mentioning as an example the oxidation of organic matter (sugars or organic acids) as source of energy for mammals or for bacteria parallel to the reduction of oxygen, nitrates, Fe (III) or sulphates. The redox potential can be seen as an indication of the kind, strength and abundance of electron donors in any system of interest, with those mentioned above known as the most prominent. The reduction of O_2 to H_2O (13.75), NO_3^- to N_2 (12.65), Fe(III) to Fe(II) (−0.8) or SO_4^{2-} to S^{2-} (−3.5) are the predominant candidates for many oxidation reactions with the number given in parenthesis [$pe°(W)$] [1] indicating the electron activity of the half redox reactions at pH 7 (written as a reduction with 1 electron accepted). The parameter *pe* is directly related to the equilibrium constant and the Gibbs free energy, indicating that the reduction of O_2 is energetically favourable compared to the others.

Equilibrium speciation calculations have been carried out for a variety of biosphere and geosphere redox sequences [2] constructing a theoretical basis that allows simulation of the behaviour of dissolved species (i.e. of metals and radionuclides) and to place their phase transitions within the redox sequence (Fig. 14.1). These theoretical redox sequences have thus the double advantage of

---

A. Albrecht (✉)
Andra, 1-7, rue Jean-Monnet, 92298 Châtenay-Malabry, France
e-mail: achim.albrecht@andra.fr

A. Bertron
LMDC-INSA, Laboratoire Matériaux et Durabilité des Constructions, Université de Toulouse, UPS, INSA, 135, avenue de Rangueil, 31 077 Toulouse Cedex 04, France

M. Libert
Commissariat à l'Energie Atomique et aux Energies Alternatives, DEN/CAD/DTN/SMTM, Saint-Paul-lez-Durance, France

F. Bart et al. (eds.), *Cement-Based Materials for Nuclear Waste Storage*,
DOI 10.1007/978-1-4614-3445-0_14,

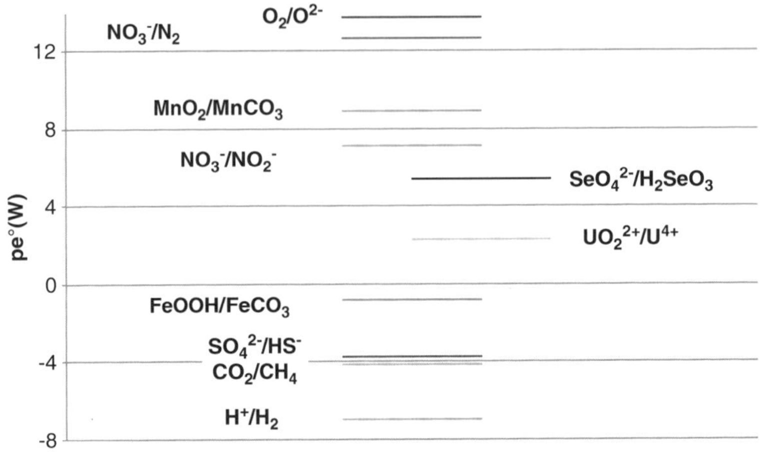

Fig. 14.1 Possible sequence of the redox potential (*pe°(W)* given for natural water at neutral pH and unit activities [1]) from an oxidising system (low electron activity, high *pe*) characterised by free oxygen, nitrates, etc., toward a reducing system (high electron activity, low *pe*) dominated by the reduction products of sulphates and carbonates; shown as examples are the potentials for the transition of Se(VI) to Se(IV) and U(VI) to U(IV)

illustrating transitions from systems initially dominated by oxidising substances, such as a nuclear waste storage cell during exploitation to a system controlled by reducing phases such as sulphides, H_2 or CH_4 and of indicating the redox potential at which a radionuclide such as Se(VI) is reduced to Se(IV), the latter known to be less soluble and more strongly adsorbing. This is a very powerful and validated means to simulate on first thermodynamic principles the speciation and the migration behaviour of any radionuclide in any system of interest for which the thermodynamic equilibrium constants are known [3].

Thermodynamics is based on equilibrium assumptions, which are rarely achieved in natural systems. This is, among others, a consequence of very slow reaction kinetics and the dependence of many redox reactions on catalysis. Nitrate, sulphate and carbonate at natural temperatures and pressures are known not to react unless catalysed by microbial activity, with some possible exceptions of abiotic surface catalysis for nitrates [4]. As these oxyanions, together with the major reducing agents organic matter, zero-valent metals and hydrogen gas play a key role in defining the redox conditions in a waste storage environment, their importance in nuclear industry, for waste stabilisation and cell construction will be discussed in a separate section.

2 Oxyanions in the Nuclear Industry

Oxyanions, in particular nitrates, sulphates and phosphates, are widely used in the nuclear industry, from fuel pellet preparation to spent nuclear fuel recycling and waste effluent water treatment. During fuel preparation, U is initially present as a uranyl nitrate solution. A variety of organic substances such as amines and urea are used to produce the fuel gel spheres, leaving behind an ammonia solution rich in nitrate and amine [5]. One main step of the recycling procedure, known as Plutonium–Uranium Extraction (PUREX) is the dissolution of the fuel rods in hot nitric acid media followed by selective solvent extraction based on tibutyl phosphate to give an example [6]. The electrochemical behaviour of nitrate can be used to introduce its reaction behaviour in purely chemical systems compared to the biogeochemical environment. The former is based on auto- [7] or surface- [8] catalytic processes that have been observed so far only at temperatures above 90 °C. The latter has extensively been studied in natural systems and requires microbial catalysis and/or the presence of reactive iron-bearing phases [9, 10].

In waste effluent treatment a variety of oxyanions are added, in particular sulphate and nitrate in acid or salt form for co-precipitation of radionuclides. To give some examples: addition of $Ti(SO_4)_2$ or $Ba(NO_3)_2$ triggers precipitation of titanium [$Ti(OH)_4$] and barium ($BaSO_4$), respectively, the former along with Sb, the latter with Sr; addition of $CoSO_4$ induces precipitation of cobalt (CoS) together with Ru [11].

In all cases effluents and precipitations are formed that contain nitrates, sulphates and a variety of organic substances. The treatment of nitrate and organic matter containing waste is difficult [5]; electrochemical reduction of nitrate (and nitrite) to nitrogen gas has been tried but the reaction demands significant energy consumption [12]. Nitrate and sulphate salt mixtures have in some countries such as Belgium, Japan or France been stabilised in a bitumen matrix. The oxyanions in such waste matrices thus often coexist with a variety of organic substances, making this waste biochemically more complex and one of the best examples to discuss the biogeochemistry of redox reactions in waste cells. It will be used here together with examples of reduced complexity.

3 Concrete, Steel, Organic Matter and Hydrogen Gas and the Storage of Nuclear Waste

Let us use the latest concept of an Andra waste cell architecture for bituminous waste to illustrate the various components present in the system (Fig. 14.2). We will not describe the cell in much detail, this has been done elsewhere [13], but simply give a rough overall understanding of volumes and components involved. The waste cell is situated within the roughly 140 m thick Callovo-Oxfordian (Jurassic) clay rock formation. The main redox sensitive components in the rock are ferric and ferrous Fe, organic matter as well as sulphates and sulphides present in celestine and pyrite; these sulphur species are currently controlling the redox potential (*pe*) which is at an equilibrium value of around −3 [14]. The waste, which is a mixture of inorganic salts embedded in a bituminous matrix, is contained in metal packages which are grouped within steel-reinforced overpacks. Here the main redox-sensitive components are the metals of the different types of steel, the organic substances of the bitumen and the oxyanions present in the waste. A large volume

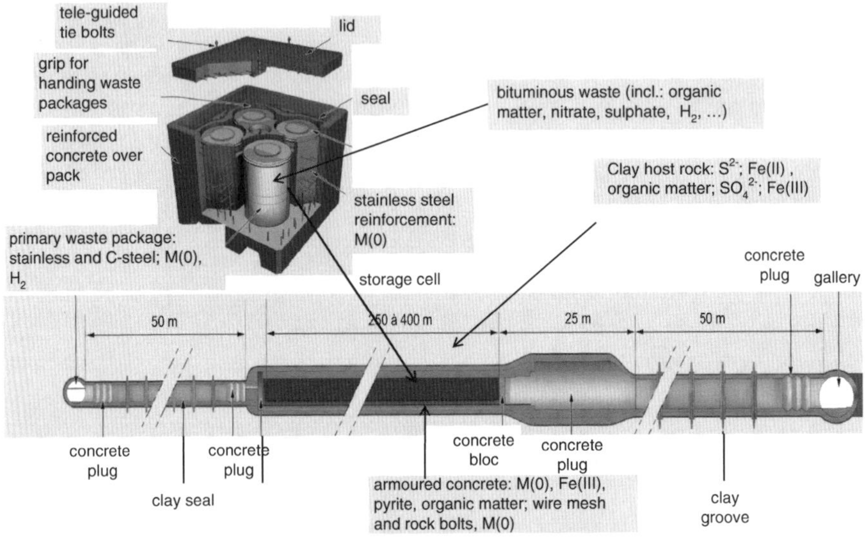

Fig. 14.2 Illustration of a repository cell for medium-level long-lived nuclear waste with the main architectural features and possible redox-sensitive components. Of particular importance here is the 250–400 m long storage cell in which 16 individual waste containers (inset) are stocked per row

of the several hundred metre long storage cell will be filled with concrete (grey zone in Fig. 14.2), bearing only few redox-sensitive components, with the possible exception of Fe(III) in hydrogarnet or Fe(II) and S(-I) in pyrite or organic matter (i.e. plasticisers). This mass of concrete will impose an alkaline condition on the cell (pH $>$ 12), with the exception of the interface zones to the primary waste containers and to the Callovo-Oxfordian host rock.

For completeness it is important to mention the steel wire mesh and the rock bolts present at the concrete–host rock interface. It will also be at this interface that the oxygen perturbation will have the most profound impact. Depending on the period of exploitation several phase changes, such as oxidation of S(-I) or Fe(II) from pyrite will occur, thus locally increasing the amount of electron acceptors [i.e. Fe(III) and sulphate]. Hydrogen gas will already be present in the waste cell during storage (produced by radiolysis of the organic matter). After waste cell closure and the disappearance of free O_2, H_2 will be produced via anaerobic corrosion, where H^+ of water serves as electron acceptor during the oxidation of Fe (0) and the other metals present in the steel. It is known that bacteria have a profound impact on anaerobic corrosion [15], but these issues will not be discussed here. During exploitation and ventilation, both the concrete as well as a defined zone of the host rock will be de-saturated. Re-saturation after closure will be a slow process, leaving the waste cell in a non-saturated state for an extensive period of time. Not shown on Fig. 14.2 is an altered zone near the interface concrete-clay rock, where fracturing can be observed as a consequence of excavation and where porosity and permeability are enhanced compared to the unperturbed host rock.

4 Microbial Activity in the Waste Disposal Cell: Probability and Impact

The high alkalinity and the reduced presence of free water (high suction, low activity) are likely to be the main parameters controlling microbial activity in a disposal cell for low and intermediate nuclear waste. A third parameter is space, but this may only be relevant for highly compacted unfractured clay rocks. The discovery of Archaea and the extremophile species associated to this group allowed expanding the domain of microbial activity to a much wider domain. The presence of bacteria living sustainably in water saturated fracture zones several kilometres below ground in South Africa [16] is of particular interest, because it shows their survival in systems where space, water and nutrition are strongly limited and where ecological exchange is remarkably hampered. The main source of energy is H_2, produced via radiolysis of water fuelled by the presence of natural radioactivity. It is also via the study of extremophile species that their capability of survival at high radioactive dose rates has been discovered and linked to a gene that helps these bacteria to survive at high suction pressure in the case of a lack of available water [17].

The final possibly limiting factor would be the high pH, but here again recent studies give more and more evidence of microbial activity even above pH 12. It is clear that the cellular protection and the required active transport mechanisms across the cell have a high energy demand and that these bacteria (alkalophiles) are characterised by strongly reduced metabolism compared to bacteria living in pH neutral conditions (neutrophiles) [18]. In a companion paper we confirm these findings for the nuclear waste-relevant case of nitrate reduction in a concrete environment [19].

It is now widely accepted that significant metabolic microbial life is not possible in highly compacted rocks such as clay host rocks (i.e. Callovo-Oxfordian). There is indirect evidence based on sulphur species evolution during compaction [20] and direct evidence from microbiological analysis of compacted bentonites [21, 22]. No such analysis has been carried out for compacted concrete, but the existence of open spaces within and between the containers and along interfaces (Fig. 14.2) does not require explicit consideration of the space argument in our context. But it is possible to limit the field of microbial activity to the near field and the excavation damaged zone of the host rock. Farther inside the rock, where no artificial fracturing occurs, microbial activity can be considered irrelevant.

The possible impact of microbial colonisation, particularly in deep geological repositories, has been considered for many years [23]. It was considered that they could not go against thermodynamics and that they would simply control the time it would take to reach the equilibrium that was used anyhow to predict the (geo-) chemistry of the repository environment. With the exception of free living microbes possibly functioning as an additional transport vector, or biofilms as an extra means of retardation, no remarkable impact was prognosticated (excluding bio-corrosion not evaluated here). Their impact on reactions that would otherwise not occur had only been discussed by few (i.e. methane production and release of ^{14}C and tritium [24]). In a more recent article [25], the positive impact of micro-organisms on the biotransformation of recalcitrant, complexing organic matter has been investigated (discussed in more detail later) and it was here that their positive impact was explicitly expressed. Their crucial impact on catalysing key reactions involving the components mentioned above (oxyanions, H_2, CO_2, CH_4, etc.) has been known [51] but not explicitly considered in the phenomenological evolution of waste cell chemistry.

5 Analogues and Applications

In this section we will on one hand consider the intermediate-level long-lived waste discussed before, and on the other look into surface disposal of low-level short-lived waste. Some bibliographic research has been done in the search for natural analogues, but the ideal case with widely comparable phenomenology has not been found. The hyper-alkaline spring waters of Maqarin in Jordan are particularly relevant, because the pH of some of the springs surpasses 12 and the site has been studied both as a natural analogue of concrete–clay interaction [26], for the migration of radionuclide analogues (i.e. Re [27]) as well as a microbiological laboratory [28]. The latter study found significant microbiological activity even at $pH > 12$, but no integrated assessment has been carried out that looked into the combined effect of microbial reaction kinetics, redox dynamics, concrete chemistry and metal mobility.

5.1 C-14

Applications of microbiological reaction catalysis can be quite striking in the case of carbon (or ^{14}C), which can be a solid (C^0 in graphite), in gaseous form of widely different valence ($C^{IV}O_2$, $C^{-IV}H_4$) or present as a variety of solid or dissolved organic substances. Many of these chemical forms have widely varying migration behaviour. If ^{14}C is present in the inorganic form (CO_2, HCO_3^-) it is known to exchange isotopically with carbonates or to be bound during carbonation of concrete, thus explaining significant retardation [29]. If the redox potential drops significantly, carbonate will become an electron acceptor (Fig. 14.1) and in the presence of methanogenic bacteria [30] ^{14}C can be converted to gaseous CH_4, known to be more mobile [24].

5.2 Tritium

The situation for tritium is somewhat comparable to ^{14}C. Tritium could also be mobilised via methane formation [24], but this pathway is less probable, because tritium is only of concern in surface or near-surface repositories (cf. short half-life) where redox conditions are less likely to be in the range of methanogenesis. A more likely scenario is the oxidation of tritiated hydrogen gas (HT), released directly from the waste containers and transported multiphase in the gaseous and dissolved forms. As titrated hydrogen gas is much less capable of entering the food chain compared to water or organically bound tritium [31], the hydrogen gas will have to be oxidised first, a reaction again requiring microbial catalysis at ambient temperature, a further example for the application of bacteria-catalysed redox reactions with an impact on radionuclide mobility and repository safety assessment. We suspect that this oxidation is less likely to take place in the concrete repository because of the high pH and the low degree of saturation, but to be preferentially occurring in the boundary zone between the repository and the underlying geosphere, possibly in the vadose zone, characterised by fluctuating groundwater levels. A variety of hydrogen-oxidising bacteria are known to occur in this zone [32], but unfortunately research efforts have mostly been focused on tritium upward flux from contaminated groundwater [33], but not the case of tritium downward movement and in all cases mostly without considering the microbial catalysis of hydrogen phase changes.

5.3 Metallic Radionuclides

We do not intent to review information on the catalysing impact of microbial activities on redox reactions implying radionuclides. Even if the radionuclide is not itself redox sensitive it is likely to be influenced by phase changes of organic matter or Fe- and Mn-(hydr)oxides, the reactivity of which is known to be mediated by bacteria [34]. We will only give a few highlights for Selenium [35] and Uranium [36], two radionuclides of relevance in safety assessments of both near-surface and deep repositories. This is done for near-neutral rather than concrete-dominated systems because of a lack of information for the latter. Both radionuclides are redox sensitive with the most oxidised form (Se^{VI} and U^{VI}) known to have a higher solubility and reduced sorption. Within the waste they are more likely present in their mobile oxidised form; their reduction is strongly catalysed by microbial activity. Even in the presence of evidence for the lack of reduction in sterile control experiments ([37] for Se; [38] for U) it cannot be precluded that very slow surface-catalysed reduction is acting. But it is important that the overall redox potential of the waste cell has reached the value where reduction is thermodynamically favourable (Fig. 14.1). If, for example, nitrate is still present in the system, the redox potential is increased and U or Se present in their oxidised form will not be reduced, even in the presence of catalysing bacteria [39]. In the case of low and intermediate waste disposal assessments, where corroding steel does not have a safety function, the positive catalytic impact of bacteria should be pointed out [40].

5.4 Organic Matter

Organic matter is both a carbon, energy and electron source for microbial metabolism. One can make a distinction between readily bio-available and more recalcitrant organic matter, with humic and fulvic acids being prime natural examples of the latter. Because they are soluble or present in colloidal form and known to carry functional groups capable of sorbing radionuclides, they have been suspected as serving as transport vectors for radionuclides that would usually be efficiently retarded by sorbing to the solid matter [41, 42]. For host rocks, such as very low porosity clay rocks, the impact of the relatively large humic and fulvic acids is less likely because of size restrictions. Waste component–degradation products such as isosaccharinic acid (ISA, degradation product of cellulose), which have a much smaller size and are known for example to increase the solubility of Pu [25] are more likely to serve as radionuclide transport vectors. In any case it has been shown that both humic and fulvic acids as well as ISA can be oxidised by microbial activity. In the case of humic substance bacteria and fungi can break the aliphatic structure either using the *N*-bearing peripheral parts or ether groups. As this requires extra energy, such activity is only found when more easily available C and N sources are not present [43, 44]. Similarly organisms capable to degrade ISA under aerobic and anaerobic conditions have been found at alkaline sites [25]. It remains unclear if the reaction goes to completeness or if intermediate—still complexing molecules—remain.

6 Conclusion and Possible Safety Approach

It is not possible to predict the evolution in space and time of the various microbial species capable of influencing key processes occurring in concrete-dominated repository systems. It is thus not really possible to predict reaction kinetics controlled by microbial activity. Two approaches are nonetheless useful; a biogeochemical simulation exercise will help tracing the reactionary paths and a mass balance approach reducing uncertainties in regard to the final, possibly equilibrium situation. Both are described here with the goal in mind to syntheses and conclude a subject matter which is at full scientific swing.

The example of a simulation exercise is for a bituminous waste cell, where nitrate and sulphate are released from the waste container, Fe^{III} is present in the system produced via corrosion and organic matter and H_2 are available as electron donors. The exercise was carried out by the National Nuclear Laboratory on behalf of Andra [45] with the biogeochemical reaction and transport code GRM [46]. The variation of the redox potential with time (Fig. 14.3) is given to illustrate the simulation approach and some major results.

Results at four distinct locations are presented, one near the waste container, the second within the concrete engineered barrier and two in the Callovo-Oxfordian clay rock, distinguishing a fractured zone from the undisturbed rock (Fig. 14.3).

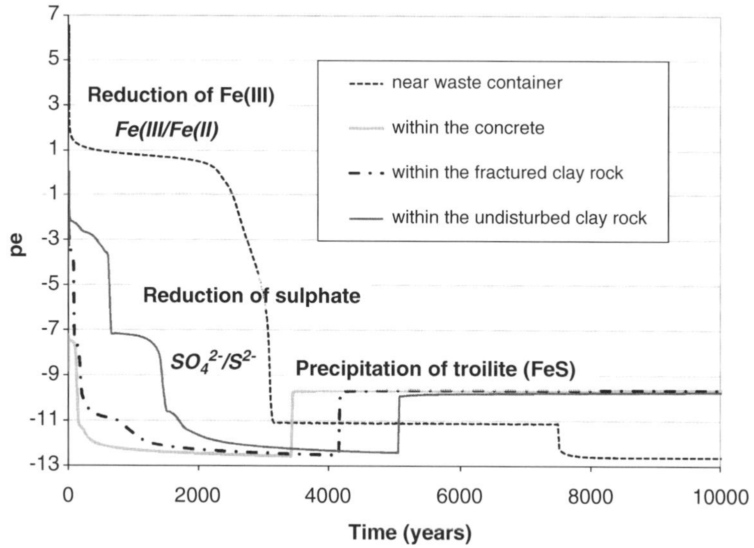

Fig. 14.3 Simulation of the redox potential in a bituminous waste cell using the biogeochemical reaction and transport (here diffusion) code Generalised Repository Model (GRM); main reactions controlling redox potential are reduction of Fe^{III} and sulphate and the precipitation of troilite. Nitrate is not considered a redox-controlling reaction

In all four locations a drop in redox potential can be seen as a consequence of reduction of Fe^{III} and sulphate, reacting either with organic matter or H_2. A variety of additional chemical reactions are considered by the code of which the formation of FeS or the degradation of concrete minerals is sufficiently important to be explicitly considered here. We are well aware that the results are a function of kinetic parameters, themselves controlled in the code on the basis of explicit consideration of growth and death rates of specific groups of bacteria. There is little known on bacterial growth in a bituminous waste cell; parameters are those used for a validation exercise of the code [46].

A much simpler approach is to carry out mass balance calculations assuming that microbial catalysis is present for all possible reactions. We will again use the example of a bituminised waste storage cell, considering for simplicity only the reduction of nitrate to nitrogen gas

$$2NO_3^- + 5H_2 + 2H^+ \rightarrow N_2 + 6H_2O$$

and sulphate to hydrogen sulphide

$$SO_4{}^{2-} + 4H_2 + 2H^+ \rightarrow H_2S + 4H_2O$$

In both cases hydrogen gas is used as an electron donor, formed via the classical anaerobic corrosion reaction

$$3Fe + 4H_2O \rightarrow Fe_3O_4 + 4H_2$$

All masses used in the mass balance exercise are based on the components present in a 250 m waste cell design [47] and the related bituminous waste inventory [48]. Corroding the total mass of steel (supposed to be 1.1×10^6 kg of Fe or 2.0×10^7 mol) would yield 2.6×10^7 mol of H_2. To reduce the 1.2×10^5 kg of nitrate (2.0×10^6 mol) requires 5.0×10^6 mol of H_2; for the 2.6×10^5 kg of sulphate (2.70×10^6 mol) another 1.1×10^7 mol H_2 are needed. The remaining amount of H_2 ($2.6 \times 10^7 - 5.0 \times 10^6 - 1.1 \times 10^7 =$) 1.0×10^7 mol will likely be used by autotrophic, methanogenetic bacteria, that will produce CH_4 by reduction of carbonates. The system could therefore run through the entire sequence as shown in Fig. 14.1. The mass of O_2 present in the waste cell after closure and the Fe^{III} initially present and produced during corrosion have been omitted in this mass balance calculation for simplicity, but they are unlikely to change the overall conclusion in particular because the masses of organic matter, second main electron donor, have also not been considered.

Knowing that bacteria control most, if not all, of the above-mentioned reactions and considering that most, if not all, of low and intermediate-level waste repositories are characterised by the massive presence of concrete, it will be important to reinforce research of microbial activity in such concrete-dominated conditions (see companion papers [19, 49]). This will have to consider not only microbial activity at high pH, but also the formation of low pH zones within concrete structures or at their interfaces as well as processes reducing their alkalinity, such as carbonation and acid–base reactions [50].

References

1. Stumm W, Morgan JJ (1996) Aquatic chemistry; chemical equilibria and rates in natural waters. Wiley, New York
2. Turner DR, Whitfield M, Dickson AG (1981) The equilibrium speciation of dissolved components in freshwater and seawater at 25°C and at 1 atm pressure. Geochim Cosmochim Acta 45:855
3. Grenthe I, Stumm W, Laaksoharju M, Nilsson A-C, Wikberg P (1992) Redox potentials and redox reactions in deep groundwater systems. Chem Geol 98:131–150
4. Postma D (1990) Kinetics of nitrate reduction by detrital Fe(II)-silicates. Geochim Cosmochim Acta 54:903–908
5. Kumar A, Rao TVV, Mukerjee SK, Vaidya VN (2006) Recycling of chemicals from alkaline waste generated during preparation of UO3 microspheres by sol–gel process. J Nucl Mater 350:254–263
6. Nikitenko SI, Venault L, Pflieger R, Chave T, Bisel I, Moisy P (2010) Potential applications of sonochemistry in spent nuclear fuel reprocessing: a short review. Ultrason Sonochem 17:1033–1040

7. Balbaud F, Fauvet P, Robin R (2010) La corrosion en milieu nitrique concentré. In: CEA-DEN (ed) La corrosion et l'altération des matériaux du nucléaire, Editions Le Moniteur, Saclay, pp 103–106
8. Truche L, Berger G (2010) Etude expérimentale de la réduction des nitrates en présence d'hydrogène et de trois différents types d'acier: acier carbone, inox 316L et Hastelloy C276; External technical document. Report number: D. RP.FSTR.1 0.0003, LMTG–Andra, Châtenay-Malabry
9. Biradar PM, Dhamole PB, Nair RR, Roy SB, Satpati SK, D'Souza SF, Lele SS, Pandit AB (2008) Long-term stability of biological denitrification process for high strength nitrate removal from wastewater of uranium industry. Environ Prog 27:365–372
10. Burgin AJ, Hamilton SK (2007) Have we overemphasized the role of denitrification in aquatic ecosystems? A review of nitrate removal pathways. Front Ecol Environ 5:89–96
11. K'Zhero R (1997) Boues bituminées: synthese des connaissances sur la coprécipitation des radionucléides sur la coprécipitation des radionucléides au cours du traitement des effluents liquides; Rapport. Report number: C.RP.AMAT.97-084, Andra, Châtenay-Malabry
12. Katsounaros I, Dortsiou M, Kyriacou G (2009) Electrochemical reduction of nitrate and nitrite in simulated liquid nuclear wastes. J Hazard Mater 171:323–327
13. Bauer C, Londe L (2008) Conception, construction et fermeture d'alvéoles de stockage MAVL; Dossier 2009. Report number: C.NSY.ASTE.08.0166, Andra, Châtenay-Malabry
14. Gaucher E, Blanc P, Bardot F, Braibant G, Buschaert S, Crouzet C, Gautier A, Girard J-P, Jacquot E, Lassin A, Négrel G, Tournassat C, Vinsot A, Altmann S (2006) Modelling the porewater chemistry of the Callovian-Oxfordian formation at a regional scale. Compt Rendus Geosci 29:55–77
15. Dinh HT, Kuever J, Mussmann M, Hassel AW, Stratmann M, Widdel F (2004) Iron corrosion by novel anaerobic microorganisms. Nature 427:829–832
16. Lin L-H, Wang P-L, Rumble D, Lippmann-Pipke J, Boice E, Pratt LM, Lollar BS, Brodie EL, Hazen TC, Andersen GL, DeSantis TZ, Moser DP, Kershaw D, Onstott TC (2006) Long-term sustainability of a high-energy, low-diversity crustal biome. Science 314:479–482
17. Rainey FA, Ray K, Ferreira M, Gatz BZ, Nobre MF, Bagaley D, Rash BA, Park M-J, Earl AM, Shank NC, Small AM, Henk MC, Battista JR, Kämpfer P, Costa MSd (2005) Extensive diversity of ionizing-radiation-resistant bacteria recovered from Sonoran Desert soil and description of nine new species of the genus Deinococcus obtained from a single soil sample. Appl Environ Microbiol 71:5225–5235
18. Horikoshi K (1999) Alkaliphiles: some applications of their products for biotechnology. Microbiol Mol Biol Rev 63:735–750
19. Libert M, Pointeau I, Sellier, R (2011) Bactéries dénitrifiantes en milieu alcalin; Final report DEN/DTN/SMTM/LMTE/NT/2011/19 - CNTPSTR1 20008, CEA Cadarache
20. Lerouge C, Grangeon S, Gaucher EC, Tournassat C, Agrinier P, Guerrot C, Widory D, Fléhoc C, Willea G, Ramboz C, Vinsot A, Buschaert S (2011) Mineralogical and isotopic record of biotic and abiotic diagenesis of the Callovian–Oxfordian clayey formation of Bure (France). Geochim Cosmochim Acta 75:2633–2663
21. Perdrial JN, Warr LN, Perdrial N, Lett M-C, Elsass F (2009) Interaction between smectite and bacteria: implications for bentonite as backfill material in the disposal of nuclear waste. Chem Geol 264:281–294
22. Stroes-Gascoyne S, Pedersen K, Haveman SA, Daumas S, Hamon CJ, Arlinger J, Ekendahl S, Hallbeck L, Gahroni N, Delaney TL, Dekeyser K (1997) Occurrence and identification of microorganisms in compacted clay-based buffer material designed for use in a nuclear fuel waste disposal vault. Can J Microbiol 43:1133–1146
23. OECD – NEA (1985) The effects of natural organic compounds and of microorganisms on radionuclide transport, NEA Workshop. Radioactive Waste Management Committee, Paris
24. Francis AJ, Dobbs S, Doering RF (1980) Biogenesis of tritiated and carbon-14 methane from low-level radioactive waste. Nucl Chem Waste Manag 1:153–159

25. Askarieh MM, Chambers AV, Daniel FBD, FitzGerald PL, Holtom GJ, Pilkington NJ, Reesb JH (2000) The chemical and microbial degradation of cellulose in the near field of a repository for radioactive wastes. Waste Manage 20:93–106
26. Gaucher EC, Blanc P (2006) Cement/clay interactions – a review: experiments, natural analogues, and modeling. Waste Manage 26:776–788
27. Trotignon L, Rose J, Khoury H, Milodowski A, Bienvenu P, Provitina O, Mercier F, Susini J (2006) Rhenium migration at the Maqarin natural analogue site (Jordan). Radiochim Acta 9–11:755–761
28. Pedersen K, Nilsson E, Arlinger J, Hallbeck L, ONeill A (2004) Distribution, diversity and activity of microorganisms in the hyper-alkaline spring waters of Maqarin in Jordan. Extremophiles 8:151–164
29. Dayal R, Reardon EJ (1994) Carbon-14 behaviour in a cement-dominated environment: implications for spent candu resin waste disposal. Waste Manage 14:457–466
30. Kotelnikova S (2002) Microbial production and oxidation of methane in deep subsurface. Earth Sci Rev 58:367–395
31. Diabaté S, Strack S (1997) Organically bound tritium in wheat after short-term exposure to atmospheric tritium under laboratory conditions. J Environ Radioact 36:157–175
32. Lechner S, Conrad R (1997) Detection in soil of aerobic hydrogen-oxidizing bacteria related to Alcaligenes eutrophus by PCR and hybridization assays targeting the gene of the membrane-bound (NiFe) hydrogenase. FEMS Microbiol Ecol 22:193–206
33. Belot Y, Watkins BM, Edlund O, Galeriu D, Guinois G, Golubcv AV, Meurville C, Raskob W, Täschner M, Yamazawa H (2005) Upward movement of tritium from contaminated groundwaters: a numerical analysis. J Environ Radioact 84:259–270
34. Lovley DR (1991) Dissimilatory Fe(III) and Mn(IV) reduction. Microbiol Mol Biol Rev 55:259–287
35. De Cannière P, Maes A, Williams S, Bruggeman C, Beauwens T, Maes N, Cowper M (2010) Behaviour of selenium in boom clay. State-of-the-art report; Report. Report number: ER-120, SCK•CEN, Mol
36. Wall JD, Krumholz LR (2006) Uranium reduction. Annu Rev Microbiol 60:149–166
37. Chung J, Nerenberg R, Rittmann BE (2006) Bioreduction of selenate using a hydrogen-based membrane biofilm reactor. Environ Sci Technol 40:1664–1671
38. Suzuki Y, Kelly SD, Kemner KM, Banfield JF (2003) Microbial populations stimulated for hexavalent uranium reduction in uranium mine sediment. Appl Environ Microbiol 69:1337–1346
39. Oremland RS, Blum JS, Bindi AB, Dowdle PR, Herbel M, Stolz JF (1999) Simultaneous reduction of nitrate and selenate by cell suspensions of selenium-respiring bacteria. Appl Environ Microbiol 65:4385–4392
40. West JM, McKinley IG, Bateman K (2008) The microbiology of redox processes – development of a redox model; Open report. Report number: OR/08/076, British Geological Survey
41. Artinger R, Krenzler B, Schüssler W, Kim JI (1998) Effects of humic substances on the migration in a sandy aquifer: column experiments with Gorleben groundwater/sediment systems. J Contam Hydrol 35:261–275
42. Kersting AB, Efurd DW, Finnegan DL, Rokop DJ, Smith DK, Thompson JL (1999) Migration of plutonium in groundwater at the Nevada Test Site. Nature 397:56–59
43. Coates JD, Ellis DJ, Blunt-Harris EL, Gaw CV, Roden EE, Lovley DR (1998) Recovery of humic-reducing bacteria from a diversity of environments. Appl Environ Microbiol 64:1504–1509
44. Filip Z, Claus H, Dippell G (1998) Abbau von Huminstoffen durch Bodenmikroorganismen – eine Übersicht. Z Pflanzenernähr Bodenk 161:605–612
45. Small J, Steele H, Eden L (2008b) Biogeochemical modelling of a bituminous waste storage cell and impact on nitrate transport; final report. Report number: Andra C RP FSTR 08.0008, Nexia Solutions, Risley, Warrington

46. Small J, Nykyri M, Helin M, Hovi U, Sarlin T, Itävaara M (2008) Experimental and modelling investigations of the biogeochemistry of gas production from low and intermediate level radioactive waste. Appl Geochem 23:1383–1418
47. Andra (2005a) Dossier 2005 Argile – Tome Architecture et Gestion du stockage géologique. Report number: C.RP.ADP.04.0001B, Andra, Châtenay-Malabry
48. Andra (2005b) Modèle d'Inventaire de Dimensionnement (MID) – Données descriptives du colis type B2; Note technique. Report number: C.NT.AHVL.02.0114B, Andra, Châtenay-Malabry
49. Alquier M, Jacquemet N, Bertron A, Erable B, Sablayrolles C, Albasi C, Basseguy R, Escadeillas G, Strehaiano P, Vignoles M (2012) Etudes expérimentales de la réactivité des nitrates à l'interface bitume – eau cimentaire – ciment en conditions biotiques; Final Report. LGC - LMDC - LCA, Université de Toulouse
50. Magniont C, Coutand M, Bertron A, Cameleyre X, Escadeillas G, Lafforgue C, Beaufort S (2009) Test method for evaluating biologically induced degradations of cementitious materials in agroindustrial environments. Concrete in aggressive aqueous environments, performance, testing and modeling, Toulouse, pp 344–355
51. Hoehler TM (2005) Biogeochemistry of dihydrogen (H_2). In: Sigel A, Sigel H, Sigel RKO (eds) Metal ions in biological systems. Taylor & Francis, Boca Raton, FL, pp 9–48

Chapter 15
Leaching of Nuclear Waste Glass in Cement Pore Water: Effect of Calcium in Solution

S. Depierre, F. Frizon, S. Gin, and F. Angeli

1 Introduction

Glass chemical interaction with cement has been studied in various fields such as in vitrified nuclear wastes or in the building industry [1–4]. In the context of nuclear waste containment glass in a geological repository, aqueous alteration studies have been performed with glass in the presence of cement. The methodology used in these investigations is closely related to the repository concept considered [5, 6]. In the French geological repository concept, intermediate-level vitrified waste packages may be disposed of in concrete containers in a deep geological repository.

Our objective is to study the glass dissolution mechanisms and kinetics in various cement pore water compositions corresponding to different stages of cement aging. In this study, we focused on one of these compositions corresponding to equilibrium with respect to Portlandite ($pH_{25^\circ C} = 12.4$) to assess the impact of calcium at high pH on the mechanisms and kinetics.

Three main mechanisms generally occur simultaneously during glass alteration by water: ion exchange (interdiffusion), hydrolysis of the silicate network and precipitation of secondary phases [7–9]. Depending on the reaction progress, however, each mechanism in turn can control the overall glass dissolution kinetics [10]. Glass composition, pH, temperature and also the leaching solution composition are key parameters for the overall glass dissolution kinetics.

S. Depierre (✉) • S. Gin • F. Angeli
Commissariat à l'Energie Atomique et aux Energies Alternatives, DEN/MAR/DTCD/SECM, 30207 Bagnols-sur-Cèze, France
e-mail: sara.depierre@cea.fr

F. Frizon
Commissariat à l'Energie Atomique et aux Energies Alternatives, DEN/MAR/DTCD/SECM, 30207 Bagnols-sur-Cèze, France

Laboratoire d'Etude de l'Enrobage des Déchets, French Atomic Energy and Alternative Energies Commission, BP17171, 30207 Bagnols sur Cèze Cedex, France

F. Bart et al. (eds.), *Cement-Based Materials for Nuclear Waste Storage*, DOI 10.1007/978-1-4614-3445-0_15,

Glass alteration mechanisms are well known in water at neutral pH [7]. However, few studies have been reported on glass alteration in basic pH and highly ionic charged solutions such as cement pore water [5, 11–13]. It is well known that glass durability decreases at high pH values [11, 14]; however, a recent study at high reaction progress and lower pH ($pH_{90^\circ C} = 8.7$) has shown that the calcium from solution induced a strong decrease in the glass alteration rate by its participation in the condensation of the passivating reactive interface [15]. Thus the elements present in the leaching solution, here the calcium coming from the Portlandite, may have an additional effect in the glass alteration mechanisms.

The purpose of this study is to have a better understanding of the effect of calcium in solution on the mechanisms and kinetics of glass leaching. A simplified glass, containing the seven main oxides of a typical French CSD-B nuclear glass, has been studied. Previous experiments have shown that this seven-oxide glass behaves like the complex one. Alteration experiments were carried out in a Portlandite-saturated solution called S2 ($pH_{25^\circ C} = 12.4$) at various glass-surface-area-to-solution-volume (SA/V) ratios and temperatures (25 and 50°C). Two hypotheses are proposed to explain the effect of calcium at both low and high reaction progress.

2 Materials and Methods

2.1 *Glass Specimen Preparation*

The simplified glass has a chemical composition corresponding to the elemental molar ratios of a typical French CSD-B nuclear glass composition for intermediate long-lived vitrified wastes (Table 15.1). The 20–40 μm and 40–80 μm powder size fractions were obtained by milling and sieving fragments of glass rods. The powder was cleaned ultrasonically in acetone and then in deionized water. The specific surface areas were determined by the BET method using Kr absorption and were equal to 1,585 and 1,005 $cm^2\ g^{-1}$ for the 20–40 μm and 40–80 μm size fractions, respectively.

2.2 *Leaching Experiments*

Boron is a good alteration tracer as it is not retained in the alteration products [16], so it can be used to determine the alteration rates except in the initial regime for which the dissolution is expected to be congruent; silicon is more easily measurable. In this regime, the forward rate and the initial rate correspond to the dissolution rate in KOH solutions and in cement pore water, respectively. The simplified glass powder samples were submitted to two types of alteration tests designed to measure the glass alteration kinetics.

Table 15.1 Glass composition in mol%

Oxide	Glass	Oxide	Glass
SiO_2	56.7	Li_2O	4.9
B_2O_3	14.0	CaO	3.7
Na_2O	13.7	ZrO_2	1.1
Al_2O_3	5.8		

2.2.1 Initial and Forward Rate Measurement

These experiments consisted of static tests at a low SA/V ratio ($0.25\ cm^{-1}$) in order to determine the forward rate in KOH solution and the initial rate in $Ca(OH)_2$ saturated solution. They were conducted at 30°C (±1°C) and 50°C (±2°C) by vigorously stirring the solution in a PTFE container with glass powder (40–80 μm fraction) and a 450 mL KOH or $Ca(OH)_2$ solution, both with a pH initially equal to 12.4 at 30°C. Samples of 3 mL were taken with a syringe at equivalent intervals during a few hours or a few days, depending on the leaching solution. These samples were filtered to 0.45 μm and the silicon concentration was measured by UV/Visible spectrophotometry according to the blue β-silicomolybdenum method with an uncertainty around 7 %. The pH was measured at the start and at the end of the alteration tests.

2.2.2 Static Test at High Reaction Progress

This experiment consisted of a static test at a high SA/V ratio ($200\ cm^{-1}$) in which glass powder (20–40 μm fraction) was leached by a 70 mL $Ca(OH)_2$ saturated solution at a slightly elevated temperature (50 ± 2°C). The initial pH was equal to 11.6 at 50°C. This test was conducted under nitrogen atmosphere to avoid carbonation of solutions and precipitation of secondary phases. Solution aliquots were taken at regular intervals over 1 year. Each sample was filtered to 0.45 μm and acidified with 0.5 N HNO_3 before ICP-AES analysis for Si, B, Na, Li and Ca with an uncertainty range of 3–5 % depending on the elements.

2.3 Expression of Results

The normalized mass loss (NL) is used to assess the quantity of altered glass from the released mobile elements (Na, Li, B) and also to determine the retention capability of the alteration products from a low mobility element (Si):

$$\mathrm{NL}_i = \frac{C_i}{x_i \cdot (\mathrm{SA}/V)}$$

where C_i is the concentration (mg L^{-1}) of element i in the solution, x_i is the mass fraction of the element i in the glass and SA/V is the ratio (m^{-1}) between the glass surface area and the solution volume.

The glass alteration is defined as follows:

$$r = \frac{d(NL_B)}{dt}$$

where r is expressed in g m^{-2} d^{-1} and NL in g m^{-2}. As specified previously, silicon would be used to determine the alteration in the initial regime instead of boron.

2.4 Characterization of the Altered Glass Samples

The glass surface of samples altered for 6 months in $Ca(OH)_2$ solution was characterized by SEM observations [ZEISS Supra 55, 15 kV acceleration voltage, equipped with a energy dispersive X-ray spectrometer (EDS)]. Direct observations of the altered grain surface have been performed to reveal the presence of secondary phases.

XPS analysis [AXIS ultra, non-monochromatic Al Kα (1,486.6 eV)] were performed on samples altered for 1 day in saturated $Ca(OH)_2$ solution at room temperature to characterize the first few nanometers of the altered glass surface. A specific protocol was set up to minimize artifacts due to the reactivity of the glass surface. More details are reported below.

3 Results and Discussion

3.1 Initial and Forward Rate Measurement

The forward and initial rates estimated from released silicon at 30 and 50°C are indicated in Table 15.2 together with the initial leaching solution pH. During the alteration experiments, the pH remained stable. For both temperatures, the results showed that the initial rates are more than an order of magnitude lower than the forward rate in the reference medium, i.e., at the same pH in KOH solution. This result highlights the compositional effect of cement pore water, especially the calcium effect.

Forward and initial rates were determined by linear regression between the set of experimental points and between the first two experimental points in KOH solution and $Ca(OH)_2$, respectively. In fact, in the case of the saturated $Ca(OH)_2$ solution, the silicon normalized mass losses seemed to evolve as a square root function of time as shown in Fig. 15.1. According to recent findings of Ca effect at lower pH

Table 15.2 Forward and initial rates at 30 and 50°C

Leachant	pH	r_0 (Si) (g m^{-2} d^{-1})
At 30°C		
KOH	12.4	0.290
S2	12.4	0.017
At 50°C		
KOH	11.7	0.742
S2	11.7	0.018

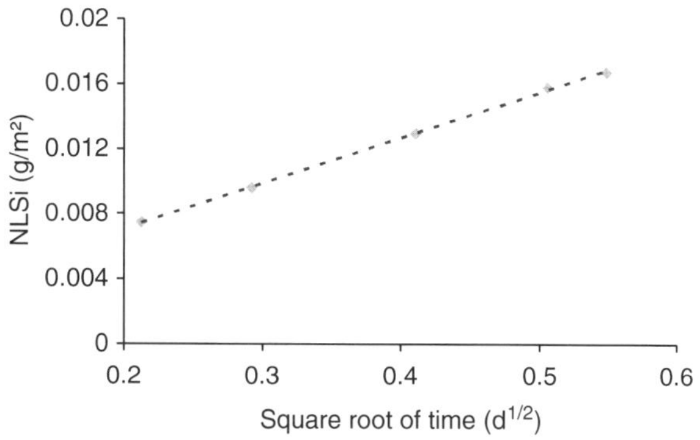

Fig. 15.1 Evolution of silicon normalized mass losses versus square root of time for the glass altered in $Ca(OH)_2$ solution at 50°C

Table 15.3 Quantitative analysis (atomic percentage) relative to different elements of simplified glass altered in saturated $Ca(OH)_2$ solution during 24 h at room temperature

	Pristine glass	Altered glass	Ratio
Si	19.6	13.8	0.7
B	7.5	5.4	0.7
Na	9.4	6.2	0.7
Al	3.8	2.1	0.5
Ca	1.3	9.8	7.3
Zr	0.4	0.3	0.7
O	58.0	62.5	1.1

Relative concentrations of the constituents are calculated excluding exogenous elements

and higher reaction progress [15], it can be assumed that the calcium coming from solution reacts with silicon at the glass/solution interface to form a passivating phase.

To further investigate this latter point, the XPS analysis was performed on glass monoliths in saturated $Ca(OH)_2$ solution (Table 15.3). It is important to note that special precautions are necessary for XPS analysis. Due to the high surface reactivity of glass with atmospheric water, XPS analysis was conducted on fresh fractured glass samples. For the altered glass, the monolith was fractured just before the beginning of the leaching experiment in a glove box to avoid contamination of the extreme surface. This analysis showed that almost all the glass components were leached at a similar ratio of 0.7 except for O and Ca. In fact, the altered glass is enriched in calcium by a factor of 7 compared to the pristine glass. The relative concentration of boron in the altered glass is 5.4 which means that either a part of the pristine glass is analyzed or that the boron is retained by the formation of a C-S-H phase on the glass/solution interface.

Geochemical calculations with the CHESS[1] code showed that as soon as the silicon concentration reaches 0.18 $\mu g\ L^{-1}$ in a Portlandite-saturated solution at 25°C, tobermorite 14-A, a natural analog of C-S-H, is potentially able to precipitate. This highlights the Si–Ca affinity at high pH. Two hypotheses can be proposed to explain the effect of calcium on the initial regime. Silicon released from the glass could react with calcium in solution to form C-S-H phases that act as a thin passivating layer on the glass surface. The calcium could also compensate two nonbridging oxygens ($Si–O^-$) in the altered layer, which could decrease the hydrolysis of silicon bonds as already shown at higher reaction progress [15]. New experiments are currently in progress to better understand this effect.

3.2 *Static Test at High Reaction Progress*

The Si, B, Na, Li, Al normalized mass losses and Ca concentration are shown in Fig. 15.2. The glass alteration rate based on the boron release of the experiment calculated by linear regression over the last 200 days is $1.4 \times 10^{-4}\ g\ m^{-2}\ d^{-1}$. The equivalent alteration thickness in boron at 270 days is 46.5 nm. The rate drop, calculated as the initial rate over the residual rate, is 375.

From the beginning of the alteration experiment, the calcium concentration decreased dramatically. After a few days, the calcium from solution was almost totally consumed and after 1 month, the calcium concentration was around 15 $mg\ L^{-1}$. The low NL(Si) indicates that silicon either recondenses to form an altered layer or participates in the precipitation of secondary phases. As indicated above, geochemical calculations showed that the pH of the solution and the calcium concentration are favorable to the formation of C-S-H phases with released silicon.

The glass powder visibly formed an aggregate at the bottom of the leaching container. Samples of altered glass powder were removed from the container after 6 months of alteration. SEM characterization was performed on the direct altered glass surface (Fig. 15.3). Numerous heaps of secondary phases can be seen on the glass grain surface, identified as C-S-H phases. As these phases are known to give mechanical properties to cement-based materials, such a precipitation can contribute to the observed cohesion between glass grains.

Complementary experiments are currently in progress to investigate the influence of the aggregation of glass powder by C-S-H phases by maintaining a constant calcium concentration using Portlandite. Our objective is to correlate the densification of C-S-H phases on the glass surface with the glass alteration kinetics.

[1] http://chess.ensmp.fr/.

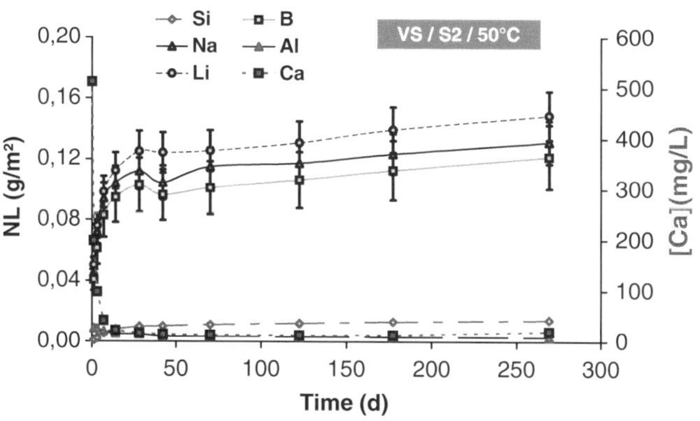

Fig. 15.2 Si, B, Na, Li, Al normalized mass losses and Ca concentration versus time in static test at SA/V = 200 cm^{-1} at 50 °C in initial saturated $Ca(OH)_2$ solution

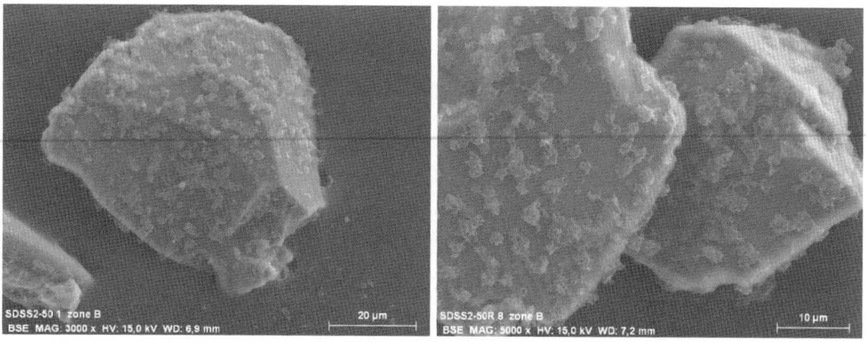

Fig. 15.3 SEM images of glass surface altered in initial saturated $Ca(OH)_2$ solution after 6 month of alteration. C-S-H phases are visible on the glass surface

4 Conclusion

The initial glass leach rate in cement pore water [$Ca(OH)_2$] depends not only on the pH and the temperature but also on the leachant composition. Calcium, which is a major component in cementitious water, diminished the high pH-dependence in highly basic media on the alteration. The calcium in solution appears to react at the glass/solution interface and limit the Si–O bond hydrolysis, perhaps by the formation of a C-S-H passivating layer at the glass surface or by inclusion of calcium in the altered layer as charge compensator of nonbridging oxygen in silicon species. At higher reaction progress, the leached silicon and the calcium in solution react rapidly forming C-S-H phases. Further experiments are in progress to study the impact of the formation of C-S-H phases on the glass alteration kinetics.

References

1. Orlowsky J, Raupach M, Cuypers H, Wastieis J (2005) Durability modelling of glass fibre reinforcement in cementitious environment. Mater Struct 38(276):155–162
2. Paul A, Youssefi A (1978) Alkaline durability of some silicate glasses containing CaO, FeO and MnO. J Mater Sci 13(1):97–107
3. Wen L, Jijian C, Yingjie H (2005) Chemical interactions between cement and E-glass fibers with a CaO–BaO–SiO2–TiO2 coating. J Am Ceram Soc 88(12):3507–3508
4. Yilmaz VT, Glasser FP (1991) Reaction of alkali-resistant glass-fibers with cement. 2. Durability in cement matrices conditioned with silica fume. Glass Technol 32(4):138–147
5. Andriambololona Z, Godon N, Vernaz E (1992) R7T7 glass alteration in the presence of mortar: effect of the cement grade. Materials research society symposium proceedings. Scientific Basis for Nuclear Waste Management Xv, vol 257, pp 151–158
6. Lemmens K, Ferrand K (2010) Report supercontainer tests on HLW glass 2004–2008 and 2009–2014, status May 2010, p 112
7. Bunker BC (1994) Molecular mechanisms for corrosion of silica and silicate-glasses. J Non Cryst Solids 179:300–308
8. Doremus RH (1975) Interdiffusion of hydrogen and alkali ions in a glass surface. J Non Cryst Solids 19:137–144
9. Ojovan MI, Pankov A, Lee WE (2006) The ion exchange phase in corrosion of nuclear waste glasses. J Nucl Mater 358(1):57–68
10. Frugier P, Gin S, Minet Y, Chave T, Bonin B, Godon N, Lartigue JE, Jollivet P, Ayral A, De Windt L, Santarini G (2008) SON68 nuclear glass dissolution kinetics: current state of knowledge and basis of the new GRAAL model. J Nucl Mater 380(1–3):8–21
11. Advocat T, Crovisier JL, Vernaz E, Ehret G, Charpentier H (1991) Hydrolysis of R7T7 nuclear waste glass in dilute media: mechanisms and rate as a function of pH. Scientific basis for nuclear waste management XIV symposium, pp 57–64
12. Ferrand K, Lemmens K (2008) Determination of the forward rate of dissolution for SON68 and PAMELA glasses in contact with alkaline solutions. Scientific Basis for Nuclear Waste Management Xxxi, vol 1107, pp 287–294
13. Gin S, Mestre JP (2001) SON 68 nuclear glass alteration kinetics between pH 7 and pH 11.5. J Nucl Mater 295(1):83–96
14. Iler RK (1979) The chemistry of silica. Wiley, New York
15. Chave T, Frugier P, Gin S, Ayral A (2011) Glass-water interphase reactivity with calcium rich solutions. Geochim Cosmochim Acta 75(15):4125–4139
16. Scheetz BE, Freeborn WPDK, Anderson C, Zolensky M, White WB (1985) The role of boron in monitoring the leaching of borosilicate glass waste forms. Mater Res Soc Symp Proc 44:129

Part IV
Emerging and Alternative Cementitious Systems

Chapter 16
Alternative Binders to Ordinary Portland Cement for Radwaste Solidification and Stabilization

C. Cau-dit-Coumes

1 Introduction

Cementitious materials intended for radioactive waste solidification and stabilization usually include substantial amounts of ordinary Portland cement (OPC) in their formulation [1]. OPC benefits from technology transfer from civil engineering research, and also from more than 150 years of experience on its durability under various service environments. However, wastes produced by nuclear activities are very diverse and some of their components may chemically react with cement phases or mixing water, reducing in some cases the quality of the product. These reactions, such as adsorption, precipitation, acid–base or redox reactions, can result in inhibition, retardation or acceleration of the cement hydration process. In some cases, the chemical interactions between waste and cement proceed slowly without affecting hydration, but eventually result in the deterioration of the waste form in storage or disposal, for instance by swelling and cracking.

The usual strategy to reduce or eliminate adverse waste–cement interactions consists in turning the penalizing constituent(s) of the waste into a form which is thermodynamically stable in cement. One alternative to avoid such a pre-treatment, which increases the complexity and cost of the process, would be using a binder showing a better chemical compatibility with the waste than OPC. A description of some common inorganic cements is given in this article, and their potential for waste stabilization/solidification is discussed. For clarity reasons, the binders are divided into three categories, depending on whether they are based on (1) hydration reactions, as for OPC, (2) acid–base reactions or (3) polycondensation reactions.

C. Cau-dit-Coumes (✉)
Commissariat à l'Energie Atomique et aux Energies Alternatives, DEN/MAR/DTCD/SPDE, 30207 Bagnols-sur-Cèze, France
e-mail: celine.cau-dit-coumes@cea.fr

F. Bart et al. (eds.), *Cement-Based Materials for Nuclear Waste Storage*,
DOI 10.1007/978-1-4614-3445-0_16,

2 Non-Portland Cement Types Undergoing Hydration Reactions: Calcium Aluminate Cements and Calcium Sulphoaluminate Cements

2.1 What Are Calcium Aluminate Cements and Calcium Sulphoaluminate Cements?

Calcium Aluminate Cements (CACs) were developed to overcome the problem associated with sulphate attack on OPC-based concrete used in the construction of railway tunnels in gypsiferous grounds [2]. The first patent was registered in 1908 by Jules Bied, from the J&A. Pavin de Lafarge company in France, and production was initiated in 1913. CACs being also characterized by rapid strength gain after setting and good resistance to abrasion compared to OPC, they are currently used in refractory and building applications, such as floor screeds and rapid-hardening mortars. Their widespread use is limited by their cost, higher than that of OPC, and directly related to the limited supply of bauxite, the main source of alumina in CAC production [3].

CACs are characterized by the presence of monocalcium aluminate (CA[1]) as their main constituent, and sometimes $C_{12}A_7$ and/or CA_2 [4]. Their chemical composition may vary over a wide range, with Al_2O_3 contents typically comprised between 37.5 and 82.5 %. Unlike Portland cement, they do not contain tricalcium silicate, but silica may be present in small amounts as dicalcium silicate (C_2S) and/or gehlenite (C_2AS). The trade names Fondu and SECAR, for cements with lower and higher Al_2O_3 contents respectively, are also widely used. They refer to the clinkering process used: either fusion of bauxite and limestone at 1,450–1,600°C for Fondu or sintering of alumina and calcium oxide (or high purity limestone) for SECAR.

The first calcium sulphoaluminate cements (CSACs) were developed in Russia and Japan in the 1960s. The change from lab scale to industrial scale was performed in China in the 1970s. Their production exceeded 1 million tons in 1999 [5], and has stayed stable around 1.2 ~ 1.3 millions tons since 2004 [6]. One of the main reasons is, as for CACs, the high cost of bauxite, one of the raw materials used.

CSACs can have highly variable compositions, but all of them contain ye'elimite, also called Klein's compound or tetracalcium trialuminate sulphate $C_4A_3\underline{S}$, in their clinker [7, 8]. In sulphoaluminate belite clinkers, ye'elimite predominates over belite, the second predominant phase [2, 9]. Other secondary phases may also be present, depending on the composition of the raw constituents [10]. In the presence of iron oxide in the raw meal, a small amount of this oxide may enter into the structure of ye'elimite, giving the solid solution $C_4A_{(3-x)}F_x\underline{S}$, with x around 0.15 [9, 11]. In addition, the ferrite phase [$C_2(A, F)$] may be formed, leading

[1] Shorthand cement notations: C = CaO, A = Al_2O_3, S = SiO_2, $\underline{S}$ = SO_3, F = Fe_2O_3, H = H_2O, T = TiO_2.

Table 16.1 Typical oxide composition (wt%) of OPC clinker, CACs and CSA clinker

		Al_2O_3	CaO	SiO_2	Fe_2O_3	SO_3
OPC clinker		3–8	59–67	17–25	0.5–6	–
CACs	Fondu	37.5–41.5	36.5–39.5	2.5–5.0	14.0–18.0	–
	SECAR 51	50.5–53	37–39	4–5.5	–	–
	SECAR 71	69.8–72.2	26.8–29.2	–	–	–
	SECAR 80	79.5–82.5	16.2–17.8	–	–	–
CSA clinker		8–47	36–61	3–27	1–13	3–15

Table 16.2 Typical phase composition of OPC, CACs and CSAC (sulphoaluminate belite cement)

		Primary phases	Secondary phases
OPC		C_3S	C_2S, C_3A, C_4AF, $C\bar{S}H_2$
CACs	Fondu	CA	$C_{12}A_7$, CA_2, C_2S, C_2AS, C_4A_F
	SECAR 51	CA	$C_{12}A_7$, C_2AS, CT
	SECAR 71	CA, CA_2	$C_{12}A_7$, A_α
	SECAR 80	CA, CA_2, A_α	$C_{12}A_7$
CSAC		$C_4A_3\bar{S}$, C_2S, $C\bar{S}/C\bar{S}H_2$	C_4A_F, C_2AS, $C_{12}A_7$, CA, CT

to the so-called ferrialuminate clinkers. The calcium aluminate phases CA and $C_{12}A_7$ may also be present if the SO_3 content in the raw meal is insufficient to convert the whole amount of Al_2O_3 into ye'elimite. On the contrary, anhydrite ($C\bar{S}$) remains in the clinker in free form at too high SO_3 contents. Free lime may also be encountered in some clinkers.

CSA clinker is produced by firing mixtures of limestone, gypsum and bauxite of appropriate compositions in shafts or rotary kilns, as for OPC clinker, but at a lower temperature (1,200–1,300°C) [12]. Industrial by-products or waste materials can be advantageously added to the blend to reduce the cost of the clinker. The clinker is then interground with calcium sulphate in much higher content (typically 15–25 wt% of gypsum [13]) than for OPC (a few wt%). By varying this content, a series of materials, ranging from rapid-hardening to shrinkage compensating, and eventually to self-stressing, can be designed [13], with applications for construction by cold weather, urgent repair, precasting, self-levelling mortars and screeds [14], glass-fibre reinforced cement composites [15], etc.

Tables 16.1 and 16.2 provide a comparison of the oxide and phase compositions of OPC, CACs and CSACs.

2.2 *How Do They Hydrate?*

Hydration of CACs and CSACs occurs according to a dissolution–precipitation process. Hydration of CA, the main anhydrous phase of CAC, can lead to four different hydrates (CAH_{10}, C_2AH_8, C_3AH_6 and AH_3) as shown in Fig. 16.1 [16]. From a thermodynamic point of view, the sole stable phase assemblage is

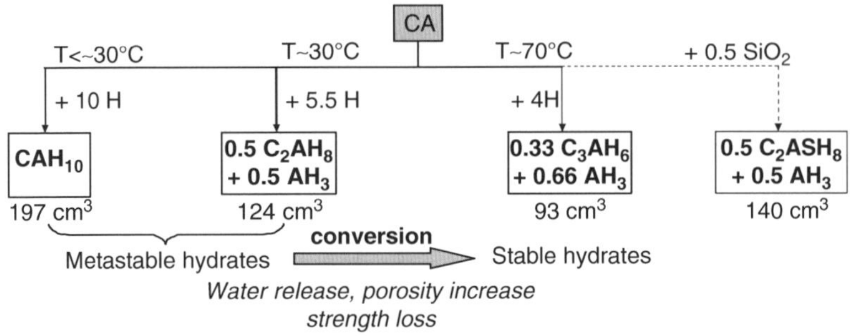

Fig. 16.1 Hydration of CA: hydrate assemblages and conversion process [16]

Table 16.3 Balance equations postulated to describe the hydration of ye'elimite [2]

Balance equation	Eq. no.
$C_4A_3\underline{S} + 18\ H \rightarrow C_3A \cdot C\underline{S} \cdot H_{12} + 2AH_3$	(16.1)
$C_4A_3\underline{S} + 2C\underline{S}H_2 + 36\,H \rightarrow C_3A \cdot 3CS \cdot 32H + 2AH_3$	(16.2)
$2C_4A_3\underline{S} + 2C\underline{S}H_2 + 52\,H \rightarrow C_3A \cdot 3C\underline{S} \cdot 32H + C_3A \cdot C\underline{S} \cdot H_{12} + 4AH_3$	(16.3)
$C_4A_3\underline{S} + 7CH + 2x\,H \rightarrow C_3AH_6 + 2C_3A \cdot {}^1\!/_2C\underline{S} \cdot {}^1\!/_2CH \cdot xH$ (approximate composition)	(16.4)
$C_4A_3\underline{S} + 8C\underline{S}H_2 + 6CH + 74\,H \rightarrow 3C_3A \cdot 3C\underline{S} \cdot 32H$	(16.5)

[C_3AH_6 + AH_3]. However, the metastable hydrates CAH_{10} and C_2AH_8 form initially at low or ambient temperature since their nucleation is easier than that of C_3AH_6. Their conversion to stable C_3AH_6 is accompanied by the formation of poorly crystallized AH_3 and the release of water. Since the density of the stable hydrates is higher than that of the metastable ones, the conversion process results in porosity increase, and subsequently in strength decrease of the material. Thus, it is the strength after conversion which must be taken into account for structural design.

When soluble silica is present, strätlingite (C_2ASH_8), another stable hydrate, may form. As for C_3AH_6, its nucleation is rather difficult, and the first precipitated hydrates are still the metastable ones. However, its density being lower than that of C_3AH_6, its formation strongly reduces the porosity increase during the conversion process. Blending CAC with silica fume [17], blastfurnace slag [18] or other sources of reactive silica [19] could thus be a way to mitigate the problem of strength loss associated with conversion.

The hydration reactions and equilibria in CSACs are complicated and not as well defined as for OPC or CAC. The hydration of ye'elimite depends on whether calcium sulphate and calcium hydroxide are also present [2, 20, 21] (Table 16.3).

- In pure water, ye'elimite is postulated to yield calcium monosulphoaluminate hydrate and aluminium hydroxide as products of hydration [Eq. (16.1)].

- Mixes of ye'elimite and gypsum produce aluminium hydroxide together with ettringite if the molar ratio between the two reactants is at least 1:2 [Eq. (16.2)], or a mixture of ettringite and calcium monosulphoaluminate hydrate if the amount of gypsum is reduced [Eq. (16.3)].
- Mixes of ye'elimite and calcium hydroxide yield hydrogarnet and AF_m phases in the absence of gypsum [Eq. (16.4)], while ettringite is the sole product if sufficient amounts of calcium hydroxide and gypsum are present [Eq. (16.5)].

Depending on the clinker composition, additional hydrates may precipitate, such as strätlingite (C_2ASH_8), C-S-H, CAH_{10} (a metastable product) or siliceous hydrogarnet. Winnefeld et al. [22] have described the phase evolution during hydration of a CSAC containing ye'elimite, belite and anhydrite with ongoing hydration. The hydration progress of the cement paste occurs by the initial precipitation of ettringite and aluminium hydroxide, followed by the precipitation of calcium monosulphoaluminate hydrate and AH_3 once calcium sulphate is deficient. Belite has a slower hydration rate and leads to strätlingite and AH_3 under the investigated conditions.

2.3 *Applications for Waste Conditioning*

The potential of CACs and CSACs for waste conditioning results at least from three factors: (1) a rate of hydration less affected by strong retarders of OPC such as heavy metals or borate ions, (2) the possibility to form in significant amounts AF_m and/or AF_t phases as well as aluminium hydroxide, which can give rise to chemical entrapment (AF_m, AF_t phases) or chemisorption (AH_3) of deleterious species of the waste and (3) a high chemical water demand, leading rather easily to dry internal environments due to self-desiccation, associated with a reduced pore solution pH compared to OPC, which could mitigate the oxidation of some electropositive metals such as aluminium.

2.3.1 A Rate of Hydration Less Affected by Some Retarders of OPC

Due to the different cement chemistries, the retarders of OPC may be less influent on CAC and CSAC hydration.

For instance, ashes resulting from the incineration of technological wastes with neoprene and polyvinylchloride may contain substantial amounts of soluble zinc chloride [23], leading to potential concentrations as high as 0.5–1 mol/L in the mixing solution. Zinc is known to have deleterious effects on OPC hydration. Setting is strongly delayed, and can even be inhibited at high zinc loadings [24], while hardening is slowed down [25]. To reduce this adverse interaction, the classical approach is to perform a chemical pre-treatment of the waste, aiming at precipitating Zn(II) as a phosphate, silicate or calcium compound. The reactions

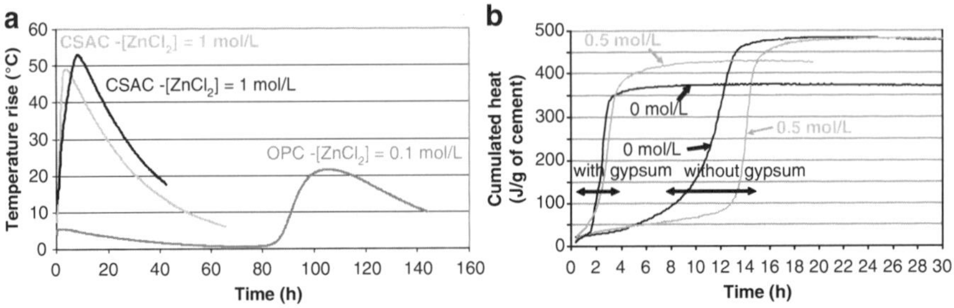

Fig. 16.2 Comparing the hydration rate of CSAC and OPC as a function of the initial $ZnCl_2$ concentration in the mixing solution. (**a**) CSA cement prepared from 80 % CSA clinker + 20 % gypsum. (**b**) CSA cement comprising either 100 % clinker or 80 % clinker + 20 % gypsum. CSA clinker comprising 68.5 wt% $C_4A_3\underline{S}$, 15.9 % C_2S, 9.5 % $C_{12}A_7$, 2.9 % CT, 1.5 % MgO, 0.5 % $C\underline{S}$, 0.5 % quartz and 2.4 % of other phases (including 1.2 % iron oxide). The curves were recorded by semi-adiabatic Langavant calorimetry. The temperature rise is characteristic of a massive dissolution of anhydrous phases. Unlike what would be expected for OPC, the small delay of the gypsum-free CSAC hydrated by the $ZnCl_2$ solution results from the strong retardation caused by chloride anions, which is partly balanced by an accelerating effect due to zinc cations. Sulphates also strongly accelerate hydration and, when gypsum is added to the binder at a level of 20 %, the delay is almost compensated [27]

with phosphates and silicates are slow at ambient temperature, and only calcium precipitation would meet industrial requirements [26]. An alternative consists in using a CSAC, which is much less retarded than OPC, as shown by Berger et al. [27] (Fig. 16.2). The waste can thus be solidified without any pre-treatment.

2.3.2 A Favourable Mineralogy for Waste Immobilization

As for OPC, the high pH of the interstitial solution of CACs (around 12.5 [28]) and CSACs (from 10.5 to 12.5 [22]) allows the precipitation of many radionuclides as hydroxides.

Moreover, the main hydrates of CSAC are ettringite and calcium monosulphoaluminate hydrate, the proportions of which depend on the calcium sulphate content of the binder: the former dominates in a gypsum-rich environment, whereas the latter tends to increase in a gypsum-deficient system. These two phases exhibit a rather flexible structure, and can accommodate many substitutions (Fig. 16.3).

CSACs have been shown to insolubilize borate ions efficiently [33]. Borates are incorporated in an AF_t phase, even with low calcium sulphate content cements. CSACs may also have a potential to stabilize waste streams with significant amounts of chloride and carbonate ions, including ^{36}Cl and ^{14}C long-lived radioactive isotopes. Mesbah et al. [34] have shown that carbonates are rapidly depleted by calcium monosulphoaluminate hydrate to form calcium monocarboaluminate hydrate and/or calcite, depending on the ratio between carbonates and monosulphoaluminate. Chloride ions react more slowly and the final chloride-containing products are Kuzel's salt and/or Friedel's salt, depending on the ratio between the two

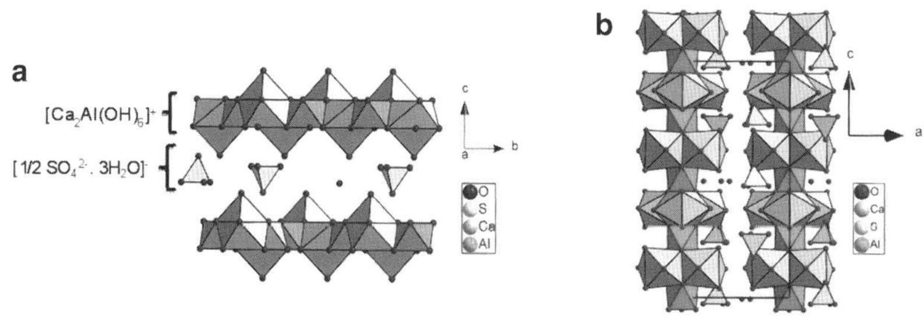

Calcium monosulphoaluminate hydrate
$C_3A.C\underline{S}.12H$

Ettringite
$C_3A.3C\underline{S}.32H$

Reported ion substitutions

Al^{3+} site	Cr^{3+}, Fe^{3+}, Ga^{3+}
SO_4^{2-} site	HBO_3^{2-}, CO_3^{2-}, Cl^-, OH^-, Br^-, I^-, NO_3^-, ClO_3^-, IO_3^-, ClO_4^-, MnO_4^-, CrO_4^{2-}, $Fe(CN)_6^{3-}$, $S_2O_3^{2-}$, WO_4^{2-}

Ca^{2+} site	Sr^{2+}, Ba^{2+}, Pb^{2+}, Cd^{2+}, Co^{2+}, Ni^{2+}, Zn^{2+}
Al^{3+} site	Cr^{3+}, Si^{4+}, Fe^{3+}, Mn^{3+}, Ni^{3+}, Co^{3+}, Ti^{3+}
SO_4^{2-} site	$B(OH)_4^-$, CO_3^{2}, OH^-, CrO_4^{2-}, AsO_4^{3-}, SeO_4^{2-}, VO_4^{3-}, BrO_3^-, NO_3^-, MoO_4^{2-}, ClO_3^-, SO_3^{2-}, IO_3^-

Fig. 16.3 Structure of calcium monosulphoaluminate hydrate (**a**) and ettringite (**b**) [29], and possible ion substitutions ([29–32] and references therein). Calcium monosulphoaluminate hydrate is an AF_m phase, belonging to the lamellar double hydroxide (LDH) large family. Its crystal structure is composed of positively charged main layers $[Ca_2Al(OH)_6]^+$ and negatively charged interlayers $[1/2\ SO_4{\cdot}nH_2O]^-$. Ettringite is an AF_t phase. Its structure is composed of positively charged columns $[Ca_3Al(OH)_6]^{3+}$ arranged parallel to the *c*-axis, and of negatively charged channels $[3/2SO_4{\cdot}nH_2O]^{3-}$

reactants. CSACs are also interesting binders for conditioning wastes with high levels of sulphates, such as demolition waste containing hydrated plaster [35]. This waste component replaces, at least partly, the calcium sulphate source usually ground with the CSA clinker, and takes part to the hydration process.

Whatever their gypsum content, CSACs have been shown to provide very good confinement of heavy metals such as lead, cadmium, zinc and trivalent chromium [36, 37]. The substitution of the heavy metal cations for calcium in the structure of ettringite is often postulated to account for the good retention. However, a recent study by Berger et al. [38] shows that, in the case of zinc, the confinement would rather result from the precipitation of a hydrated and well crystallized compound with platelet morphology, possibly a layered double hydroxide, at early age (<1 day), and from chemisorption of Zn^{2+} onto aluminium hydroxide at later age.

CACs also form AH_3 with sorption properties. Moreover, when CACs are hydrated with ionic solutions, AF_m phases can precipitate. A patent has been registered on the use of CACs for purifying aqueous solutions polluted by nitrate ions, which are precipitated as calcium nitroaluminates [39]. The nature of the precipitated hydrates can be adjusted by blending CAC with calcium sulphate (to promote sulphate-AF_m or AF_t phases precipitation), calcium hydroxide (to promote OH-AF_m phases), or fine limestone (to precipitate carbonate-AF_m phases). Toyohara et al. [40] showed for instance that the best CAC to bind radioiodide comprises SECAR71 and gypsum in weight ratio of 100:15.5. The enhanced sorption

Table 16.4 Aluminium corrosion rates in CSAC and OPC/blast furnace slag (BFS) matrix [44]

Binder	Peak rate during the first 24 h (L/h/m^2)	Rate (L/h/m^2) at 24 h	4 days	40 days
CSA1	0.22	0.010	0.0002	Below detection
CSA3	0.1	0.010	0.0023	Below detection
BFS:OPC	5.8	0.81	0.10	Not measured

CSA clinker composition: $C_4A_3\bar{S}$ 58 %, C_2AS 17 %, CA 16 %, C_4AF 4 %, CT 4 %, $C_{12}A_7$ 1 %
Cement compositions: CSA1: 60 wt% clinker/40 % gypsum at $w/c = 0.6$; CSA3: 70 % clinker/30 % gypsum at $w/c = 0.65$; BFS:OPC 90:10, $w/c = 0.33$, $T = 40$ °C

compared to OPC is attributed to the precipitation of an I-AF_m phase. Interesting results have also been reported for caesium, another labile radionuclide in hydrated OPC. By blending SECAR71 with 40 wt% silica fume, Fryda et al. [41] showed that more than 99 % of inactive caesium, can be immobilized in a zeolitic phase of the chabazite family. Moreover, the same zeolitic phase precipitates when a pre-hydrated mix of CAC and silica fume is in contact with a Cs solution.

2.3.3 Self-Desiccation and Lower Alkalinity of the Pore Solution Could Mitigate the Oxidation of Aluminium

Aluminium is a reactive amphoteric metal, readily forming a protective oxide layer on contact with air or water. This layer is generally regarded as stable in the pH range 4.5–8.5. However, in a strong alkaline solution, such as the pore solution of OPC, the layer is soluble and corrosion continues, with production of dihydrogen, formation of expansive metal hydroxides as well as calcium-based aluminosilicates [42]. Using CSAC could lead to a marked reduction in aluminium corrosion as compared with composite OPC system [43, 44] (Table 16.4).

Some corrosion is detected in the first 24 h, but, following this initial period, rates fall to very low levels. The interstitial pH of the CSAC matrix, around 11, is lower than that of the BFS/OPC reference (pH around 13), but still outside the range of passivation of aluminium, which could explain the initial corrosion. Subsequently, the binding of a large amount of water into the ettringite structure may lead to self-desiccation of the system, lowering the internal humidity, and thus reducing the ongoing corrosion. The reported studies were carried out on relatively short periods (28–40 days), and complementary investigations would be required to assess the long-term corrosion of aluminium. Indeed, the pore solution pH of a CSAC paste is between 10 and 11 at early age, but increases to about 12.5 once calcium sulphate is exhausted. The risk of corrosion restart in hydrated monoliths should be assessed in case of water resaturation of the materials.

Recently, a special grout, compatible with aluminium components, has been designed for the decommissioning of old reactors at Savannah River site (USA). It is based on CAC blended with calcium sulphate (66.6 % Fondu + 33.7 % of Plaster of Paris), which hydrates to form ettringite and develops an interstitial solution pH of ~9.5. About 92 m^3 were poured into a reactor vessel at the end of 2010 [45].

3 Phosphate Cements Based on Acid–Base Reactions

Phosphate cements are the main representatives of acid–base cements. The setting/hardening process is brought by a reaction between acid and basic compounds, yielding a salt or hydrated salt. The acid reactant may be an inorganic acid, or an acid salt (for example monoammonium phosphate, or acid phosphates and polyphosphates of ammonium and alkali metals). The basic constituent is usually a weakly basic or amphoteric metal oxide with a moderately small ionic radius (MgO, ZnO, CaO), but may also include other compounds that are less acid than their counterpart (such as tetracalcium phosphate $Ca_4(PO_4)_2O$ or wollastonite ($CaO{\cdot}SiO_2$) [2].

3.1 Magnesium Phosphate Cements

The main area of application of magnesium phosphate cements is mortars or concrete mixes for rapid repair works. One of the advantages is the good bonding of magnesium phosphate cement mixes to mature concrete based on OPC.

Their constituents are magnesium oxide (calcined, or "deadburnt", magnesia) and a water-soluble acid phosphate, which is most often diammonium hydrogen phosphate $(NH_4)_2HPO_4$. Struvite ($NH_4MgPO_4{\cdot}6H_2O$) is the main product of reaction [Eq. (16.6)], and the one mainly responsible for setting and hardening.

$$MgO + (NH_4)_2HPO_4 + 5H_2O \rightarrow NH_4MgPO_4 \cdot 6H_2O + NH_3 \tag{16.6}$$

Other by-products may also form in unwanted side reactions, such as dittmarite ($NH_4MgPO_4{\cdot}H_2O$), schertelite $(NH_4)_2Mg(HPO_4)_2{\cdot}4H_2O$, newberite $MgHPO_4{\cdot}3H_2O$ and magnesium phosphate $Mg_3(PO_4)_2{\cdot}4H_2O$. An amorphous or poorly crystalline phase may also be precipitated in mixes with very low water contents [46]. Noxious gaseous ammonia is released in the hardening process, which is not suitable for radioactive waste immobilization. This problem can be avoided by using alkali phosphates (such as NaH_2PO_4, KH_2PO_4, Na_2HPO_4, K_2HPO_4) instead of ammonium phosphate. The counterpart is a decrease in the final strength of the hardened material.

In Argonne National Laboratory (USA), a matrix called ceramicrete was developed to treat problematic low-level mixed wastes [47]. The process is based on a reaction with deadburnt MgO and monopotassium phosphate KH_2PO_4, leading to the precipitation of $MgKPO_4{\cdot}6H_2O$ (MKP). The reaction is rapid and exothermic. Boric acid (at a typical content of 1–2 wt% of the binder) has thus to be used as a retardant to get an acceptable setting time and to limit the temperature rise during setting and hardening. Langton et al. [45] noticed however that a minimum temperature of 65°C is necessary to form a significant amount of MKP. At lower temperatures, other hydrated magnesium potassium phosphate phases precipitate, with no binding power. The solidified binder typically exhibits a compressive strength between 20 and 30 MPa, a density of ~1.8 g/cm^3, an open porosity of 2–5 vol%, and a pore solution pH within the range 6–8 [48].

Waste immobilization results from two processes: precipitation of many contaminants (in particular actinides) as phosphates with very low solubility, and physical encapsulation in a dense phosphate matrix. Good results have been reported for several types of wastes: low-level debris wastes contaminated by ^{137}Cs [49], ^{99}Tc oxide-containing wastes [50], Pu-containing combustion residues [51], or highly saline effluents [52]. Stabilization of As, Cr and Hg is however difficult. These contaminants may require additional stabilizers, such as a source of sulphides, to decrease their solubility. Besides, sulphates have to be reduced before encapsulation, and immobilization of caesium requires a prior treatment with zirconium phosphate.

Because of its low pore solution pH which should limit oxidation of aluminium, ceramicrete was selected as a potential candidate for reactor vessel filling in the Savannah River in situ decommissioning project [45]. A pumpable, flowable, self-levelling slurry was formulated. Finally, it was not selected (see Sect. 2.3.3) but, because of its interesting features, such as adhesion to metal surfaces, absence of shrinkage and chemical stabilization of radionuclides, it is still under consideration for other decommissioning and dismantling applications.

3.2 Calcium Phosphate Cements

Calcium phosphate cements are composed of one or several calcium phosphates, to which other calcic compounds can possibly be added, such as calcium hydroxide or calcium carbonate, as well as additives to control the setting rate and the properties of the hardened matrix. Mixed with water, they yield stoichiometric or calcium-deficient hydroxyapatite by an acid–base or hydrolysis reaction. Their main area of application is dentistry and medicine since they form a product with a composition close to that of bones and teeth. They could also offer some prospects for nuclear waste conditioning given the outstanding properties of hydroxyapatite.

- The structure is highly tolerant to cation and anion substitution, which can lead to chemical entrapment of deleterious species (Fig. 16.4).
- Stoichiometric hydroxyapatite (Ca/P = 1.67, no vacancy on the cationic sites) is stable up to 1,000°C.
- It has the property to anneal radiation damage, which gives it low amorphization susceptibility [54].
- It exhibits a very low solubility in water, respectively three and almost four orders of magnitude lower than that of C-S-H and portlandite. Besides, it is the less soluble calcium phosphate in slightly acidic, neutral or basic medium [55] (Fig. 16.5), and its solubility is retrograde, decreasing when temperature increases ($K_s = 10^{-117}$ at 25°C, and 10^{-126} at 100°C [56]).

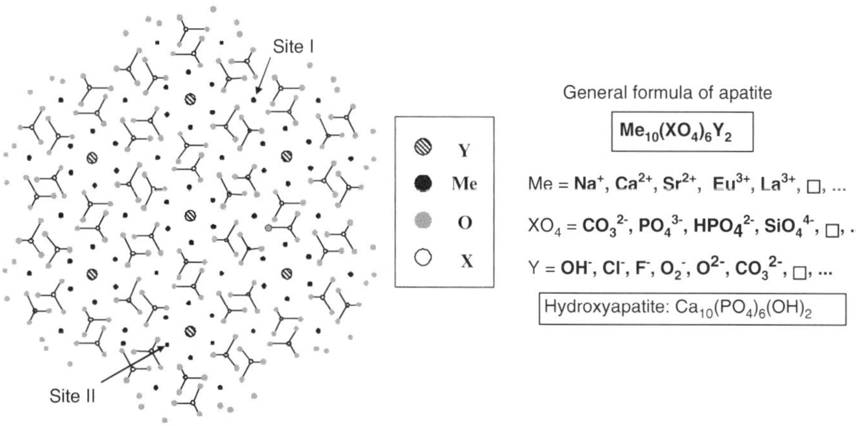

Fig. 16.4 Crystallographic structure of hydroxyapatite. Projection on the [001] plane [53]. Hydroxyapatite crystallizes in the hexagonal P_63/m space group. It is made from a quasi-compact stacking of phosphate groups which form two types of tunnels, parallel to the *c*-axis, in which the calcium ions are localized. The first type of tunnels, with a diameter around 2.5 Å, is located between three phosphate groups, and its centre corresponds to cationic site I. The second type of tunnels is larger, with a diameter between 3 and 4.5 Å, and located between six phosphate groups. Six cationic sites, referred as sites II, are located alongside these tunnels at a height of c/4 (three sites) and 3c/4 (three sites), and the centre is occupied by a hydroxide ion. The presence of vacancies (*open rectangle*) can be caused by the departure of cations or anions, or by the presence of cations or anions with different valencies. The existence of tunnels in which the anions are located gives to hydroxyapatite properties similar to zeolites. For instance, they can behave as ion exchangers (OH^-, F^-, Cl^-)

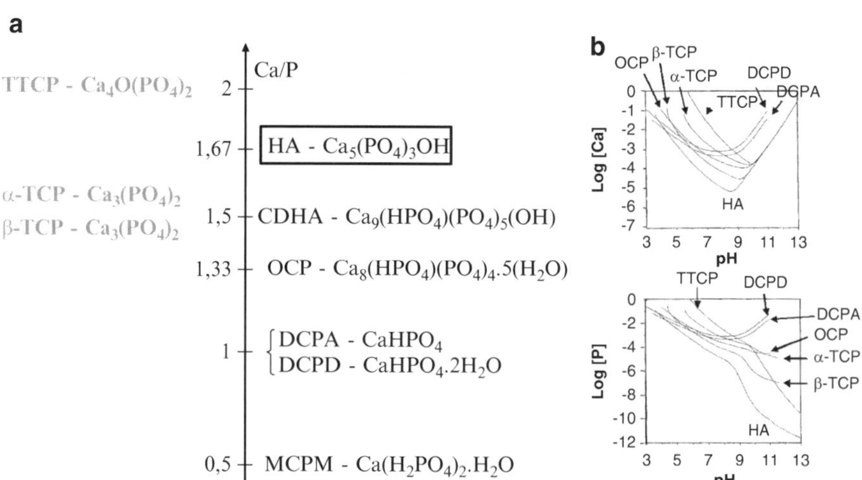

Fig. 16.5 Main calcium phosphates (**a**) and their solubility isotherms at 37 °C [55] (**b**)

Calcium phosphate cements are characterized by a high variety, resulting from the diversity of possible precursors (Fig. 16.5a). The reactions leading to stoichiometric or calcium-deficient hydroxyapatite are hydrolysis, or acid–base.

- Hydroxyapatite is the less soluble calcium phosphate for a pH higher than 4.2 (at 37°C) (Fig. 16.5b). Thus, in water suspensions, all calcium phosphates have the potential to convert to hydroxyapatite according to a dissolution–precipitation process.
- Tetracalcium phosphate is the only calcium phosphate with a Ca/P ratio higher than that of hydroxyapatite. It can be combined with another calcium phosphate of lower Ca/P to form hydroxyapatite according to an acid–base reaction.
- Stoichiometric hydroxyapatite can also be formed from blends of calcium phosphates free from tetracalcium phosphate provided a complementary addition of calcium in the form of calcium hydroxide or calcium carbonate to adjust the Ca/P ratio to 1.67.

All these reactions do not necessarily lead to a cementitious system with reasonable setting time. The reaction rate can be accelerated by different ways:

- adding a diluted H_3PO_4 solution as the mixing solution, instead of plain water,
- varying the pH of the mixing solution, which may be done by varying the XH_2PO_4/X_2HPO_4 ratio ($X = Na^+$ or K^+),
- adding crystalline hydroxyapatite to the mix as a seeding material,
- increasing the fineness of the starting materials.

It is also particularly important to determine the water balance: if low porosity is required, most of the added water should be chemically combined, which is not always the case, some reactions even producing some water (for instance TTCP + MCPM, TTCP + DCPD or TTCP + OCP).

Most available calcium phosphate cements are not appropriate for waste conditioning since they have been developed for medical applications, with requirements very far from those expected in the nuclear field. However, six reaction schemes were compared in order to form hydroxyapatite at room temperature, with the objective to develop a cement aiming at encapsulating nuclear spent fuel assemblages [α-TCP/H_2O, α-TCP/$CaCO_3$, α-TCP/$Ca(OH)_2$, TTCP/DCPA, DCPA/$Ca(OH)_2$ and DCPA/$CaCO_3$] [57]. Hydrolysis of α-TCP and acid–base reaction between TTCP and DCPA gave the most promising results. In both cases, the product is a calcium deficient apatite. Cement pastes ($w/c = 0.4$–0.5) with setting times below 24 h are obtained provided additions of nucleation seeds in the solid phase and sodium monohydrogenophosphate in the liquid phase. Flow properties are improved by introducing a plasticizer (polycarboxylic ether). However, hardening is quite slow, and the cement pastes exhibit a compressive strength around 8 MPa only after 28 days. It results from their high porosity (around 50 %), which is itself the consequence of the low water demand of the chemical reactions.

4 Alkali-Activated Binders

4.1 What Are Alkali-Activated Binders?

Alkali-activated binders are made by mixing solid aluminosilicates, such as fly ash, metakaolin, various clays usually activated by heat or blastfurnace slag, with an activating solution comprising high concentrations of alkali hydroxide (NaOH, KOH) and/or polysilicate ($Na_2O{\cdot}n SiO_2$, $K_2O{\cdot}n SiO_2$). The reaction product, formed according to a polycondensation process, exhibits a network structure that could evoke that of organic thermoset polymers. Thus, the binders are sometimes called "inorganic polymers" [58], or "geopolymers" [59].

Geopolymer synthesis chemistry depends on the nature of the solid precursor and alkali activator, and many aspects are not fully understood. However, it is possible to describe a general and simplified process [60].

- Dissolution of the solid aluminosilicate source by alkaline hydrolysis, consuming water, produces aluminate and silicate species.
- Once in solution, the silicate and aluminate precursors are mobile and rearrange to form more stable oligomers.
- Dissolution of amorphous aluminosilicates leads to oversaturation of the solution. The oligomers form a large network by condensation, and a gel-like phase precipitates. This process releases the water that was consumed during dissolution. According to NMR and FT-IR results, the gel-like phase exhibits a high aluminium content (Si/Al ratio ~1), which can be attributed to the readier dissolution of aluminium than silicon because the Al–O bonds are weaker than the Si–O bonds.
- As the reaction progresses, more Si–O groups in the starting material dissolve, favouring the evolution of the initial gel-like phase into a more silica-rich product (Si/Al ratio ~2). The connectivity of the network increases. This process of reorganization determines the microstructure and pore distribution of the final material.

The gel-like aluminosilicate phase bears strong resemblance to zeolite framework, but is generally lacking in long-range crystalline order. Its structure consists of a rather randomly arranged three-dimensional network of corner-linked SiO_4 and AlO_4 tetrahedra (Fig. 16.6).

To outbalance the negative charges brought by the presence of AlO_4 tetrahedra, a corresponding number of positive charges, in the form of monovalent or divalent cations, are located in spaces between the tetrahedra, together with some constitutional water. This type of product is described in the literature as N-A-S-H gel. Zeolite crystallization tends to occur with high water content, high temperature curing and/or extended ageing, and is more important in hydroxide systems [62].

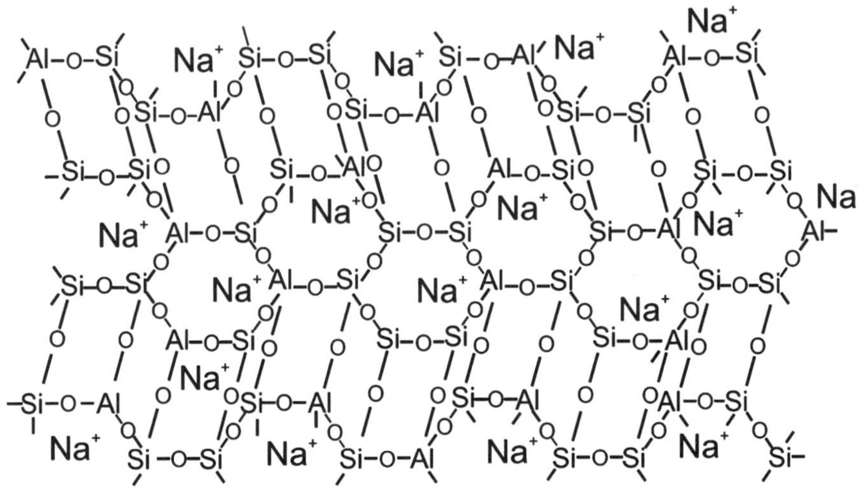

Fig. 16.6 Schematic structure of a Na-Poly(sialate-siloxo) gel according to Barbosa et al. [61]

4.2 Potential of Alkali-Activated Binders for Waste Conditioning

By selecting appropriate starting materials and by varying the conditions of processing and curing, it is possible to vary the properties of the produced alkali-activated binders over a wide range, and to tailor them to specific requirements [63, 64]. Geopolymers can thus exhibit a wide range of properties, including high compressive strength, low permeability, low shrinkage, fast or slow setting, low thermal conductivity, acid, fire and/or freeze–thaw resistance [65]. Geopolymerization may also provide an opportunity to stabilize/solidify hazardous wastes.

The mechanisms of stabilization are believed to be physical, due to the low permeability of the matrix, and/or chemical. Several processes can occur, depending on the waste contaminant, such as precipitation as a compound (typically a silicate or hydroxide) of low solubility, or binding into the structure for charge balancing roles. For instance, well-designed geopolymers have been shown to provide good immobilization of lead [66] which is believed to precipitate as Pb_3SiO_5 [67]. Ga and Ge substitute for Si or Al in the gel structure [68]. Alkali metals (Li, Na, K, Rb, Cs) are mainly immobilized by acting as charge balancing cations within the gel [69]. In particular, the insolubilization of caesium by a geopolymer (either made of fly ash or metakaolin) is reported to be much better than by an OPC matrix [70].

Bankowski et al. [71] investigated the leaching of various cationic contaminants from a mixed brown coal fly ash/metakaolin geopolymer. Every s- (Ba, Sr) or p-group (As, Se) element was effectively immobilized by the geopolymerization process, while the transition metals either showed no immobilization (Cr, Cu, Mo) or increased leachability (Mn, V, Zn). Upon exposure to very high pH, the speciation of transition metals may change to favour the formation of soluble oxyanions. Geopolymers are thus more effective as a means of treating wastes that do not significantly increase in solubility at high pH.

The high alkalinity of the interstitial solution of a geopolymer, at least at early age, is beneficial to reduce the corrosion of Mg–Zr alloys, compared to OPC. According to Lambertin et al. [72], the corrosion density of such alloys in a geopolymer activating solution is one order of magnitude lower than in an OPC pore solution. When the alloys are encapsulated in a geopolymer, the H_2 production due to corrosion is 2–3 times lower than in the OPC matrix during the first 28 days.

Very few studies concern the influence of anions on geopolymers evolution. The presence of sulphates accelerates the conversion of the N-A-S-H gel-like phase into zeolites [73]. This, in turn, can adversely affect the mechanical strength of the material. By comparison, nitrates seem less effective in favouring the geopolymer/zeolite transformation [74].

Application of geopolymers to radioactive waste conditioning is still limited. Khalil and Merz [75] investigated the use of a commercial geopolymer to immobilize intermediate-level wastes containing Cs, Mo and Sr. Static leaching tests were performed in distilled water and Q-brine ($MgCl_2$ saturated salt brine). Strontium was better immobilized than caesium, which was itself better confined than molybdenum. The matrix exhibited good resistance to leaching by pure water: the samples maintained their shape, dimensions and strength. On the contrary, tolerance of most compositions to Q-brine was poor.

More recently, Lichvar et al. [76] reported the development of an alkali-activated binder, named SIAL® matrix, for the solidification of radioactive sludge and sludge/ion exchange resin mixtures. The dry matter loading in the final products was in the range 12–17.5 wt%. The solidified waste forms checked the Czech and Slovakian specifications for disposal in a surface repository, with leachability indexes higher than 6 (according to the ANSI/ANS-16.1-1986 procedure—^{137}Cs: $L_i = 9–14$, Am, Pu nuclides: $L_i = 12–18$, ^{90}Sr: 12–14), and compressive strength about 25 MPa. In the last years, 29.6 m^3 (38.15 t) of radioactive waste with a total activity of ~4.94×10^{12} Bq were immobilized using the SIAL® matrix.

Some authors suggest that geopolymers may be viewed as amorphous analogs of zeolites or that, from a thermodynamic point of view, geopolymers can be considered as metastable with regard to zeolites [77]. Zeolites are a family of complex aluminosilicates having a three-dimensional network structure containing channels and cavities which can immobilize a variety of contaminants. Controlling the crystallization of geopolymers and tailoring the particular zeolite structures formed may be very attractive to ensure optimal immobilization of the desired contaminants. In this context, materials called hydroceramics have been developed to deal with reprocessing wastes at the Idaho National Engineering and Environmental Laboratory (INEEL, USA) [78]. INEEL waste is characterized by overwhelming amounts of sodium. The matrix constituents include waste (typically 30 % dry-mass basis), metakaolin or class C fly ash, ~5 % powdered vermiculite

Table 16.5 Concerns and key issues for the investigated alternative binders

Elaboration	Phosphate cements	Cost and availability of the binders components Calcium phosphates prepared at high temperature (TTCP, α-TCP) are expensive Specific surface area of deadburnt MgO can be highly variable from one batch to another, affecting the binder reactivity
	Geopolymers	Handling of large amounts of highly concentrated solutions Most superplasticizers used for OPC are not effective
Heat production	CACs/CSACs Phosphate cements Geopolymers	Careful management of the initial exotherm is needed
Setting and hardening	CACs/CSACs	CSACs can set very rapidly depending on their ye'elimite content, the kind and content of minor phases, and the amount and reactivity of calcium sulphate. In that case, a retarder (boric acid, citric acid) has to be used
	Phosphate cements	Setting can be too rapid for magnesium phosphate binders, or some calcium phosphate cements based on MCPM. A retarder is then needed Setting of calcium phosphate cements involving $Ca(OH)_2$ and calcium phosphates prepared at room temperature is too slow for industrial application Slow hardening of calcium phosphate cements; limited strength due to high porosity
	Geopolymers	Setting and hardening process highly sensitive to the activation conditions : sometimes, slow strength gain
Durability	CACs/CSACs	CACs weakened by a prolonged service in warm and humid environments (conversion); insufficient experience on the long-term evolution of CAC/SiO_2 blends Hot and dry environments should be avoided for CSACs (ettringite prone to lose water)
	Phosphate cements	Durability unproven (especially for magnesium phosphate cements)
	Geopolymers	Durability unproven What is the influence of the geopolymer/zeolite transformation on the properties of the matrix? Are the zeolites representative of the natural evolution of the matrix?
Interaction with wastes	CACs/CSACs Phosphate cements Geopolymers	Need for a better understanding of the processes involved

(to enhance ^{137}Cs fixation), ~0.5 % sodium sulphide (redox buffer and heavy metal precipitant), plus ~10 % sodium hydroxide dissolved in enough water to produce a stiff paste. The hydroceramic waste forms are then autoclaved at 90 or 190°C to get a dense matrix with very low solubility, consisting mainly of zeolites.

5 Conclusion

Binders with different levels of development have been presented in this article. CACs have been used for more than 100 years under various service conditions, and are much better known than more novel binders such as magnesium phosphate cements or geopolymers. Table 16.5 tries to summarize the main concerns and remaining key issues for the different investigated binders.

In all cases, the reactions involved in the setting/hardening process are exothermic. While excessive temperature rise is not a problem affecting small-scale laboratory samples, large-volume drums of cemented waste forms may exhibit a substantial temperature rise. This thermal evolution at early age should be taken into account to understand how actual cemented waste forms will perform. The question is all the more important since the solid phase composition in materials based on CSACs, CACs, magnesium phosphate and alkali-activated binders depends on temperature. Besides, understanding the chemistry of cement–waste interactions, and their consequences on the physical properties of the solidified waste forms, including their long-term evolution, will be a critical task for the acceptance of these alternative binders in nuclear waste conditioning. Particular attention will also have to be paid to their possible interactions with the near-field environment. This offers a wide field of research which should be attractive and stimulating for physico-chemists in the years to come.

References

1. Atkins M, Glasser FP (1992) Application of Portland cement-based materials to radioactive waste immobilization. Waste Manage 12:105–131
2. Odler I (2000) Special inorganic cements. Taylor & Francis Group, London
3. Juenger MCG, Winnefeld F, Provis JL, Ideker JH (2011) Advances in alternative cementitious binders. Cement Concr Res 41:1232–1243
4. George CM (1983) Industrial Aluminous cements. In: Barnes P (ed) Structure and performance of cements. Applied Science Publishers, London, pp 415–469
5. Zhang L, Su MZ, Wang YM (1999) Development of the use of sulpho- and ferroaluminate cements in China. Adv Cement Res 11:15–21
6. Wang J (2010) Hydration mechanism of cements based on low-CO_2 clinkers containing belite, ye'elimite and calcium alumino-ferrite. PhD Thesis, Lille 1 University, France
7. Klein A, Troxell GE (1958) Studies of calcium sulphoaluminate admixture for expansive cements. Proc ASTM 58:986–1008

8. Mehta RK (1965) Investigation on the products in the system $C_4A_3\underline{S}$-$CaSO_4$-CaO-H_2O. Proceedings of the annual meeting of the Highway Research Board, pp 328–352
9. Sharp JH, Lawrence CD, Yang R (1999) Calcium sulphoaluminate cements: low-energy cements, special cements or what? Adv Cement Res 11:3–13
10. Sahu S, Majling J (1993) Phase compatibility in the system CaO-SiO_2-Al_2O_3-Fe_2O_3-SO_3 referred to sulphoaluminate belite cement clinker. Cement Concr Res 23:1331–1339
11. Chen D, Feng F, Long S (1993) The influence of ferric oxide on the properties of $3CaO.3Al_2O_3.CaSO_4$. Thermochim Acta 215:157–169
12. Ikeda K (1980) Cements along the join $C_4A_3\underline{S}$-C_2S. Proceedings of the 7th international congress of the chemistry of cement, Paris, pp 31–36
13. Glasser FP, Zhang L (2001) High-performance cement matrices based on calcium sulphoaluminate-belite compositions. Cement Concr Res 31:1881–1886
14. Georgin JF, Ambroise J, Pera J, Reynouard JM (2008) Development of self-levelling screed based on calcium sulphoaluminate cement: modelling of curling due to drying. Cement Concr Compos 30:769–778
15. Pera J, Ambroise J (2004) New applications of calcium sulphoaluminate cement. Cement Concr Res 34:671–676
16. Fryda H, Saucier F, Lamberet S, Scrivener K, Guinot D (2010) La durabilité des bétons d'aluminates de calcium. In: Ollivier JP, Vichot A (eds) La durabilité des bétons, bases scientifiques pour le développement de bétons durables dans leur environnement. Presses de l'Ecole Nationale des Ponts et Chaussées, Paris, pp 767–823
17. Ding J, Fu Y, Beaudoin JJ (1995) Strätlingite formation in high alumina cement/silica fume systems: significance of sodium ions. Ceram Concr Res 25:1311–1319
18. Majumdar AJ, Edmonds RN, Singh B (1990) Hydration of SECAR 71 aluminous cement in presence of granulated blast furnace slag. Cement Concr Res 20:7–14
19. Ding J, Fu Y, Beaudoin JJ (1997) Effects of different zeolites on conversion-prevention in high alumina cement products. ACI Mater J 94:220–226
20. Hanic F, Kaprálika I, Gabrisováa A (1989) Mechanism of hydration reactions in the system $C_4A_3\underline{S}$-$C\underline{S}$-CaO-H_2O referred to hydration of sulphoaluminate cements. Cement Concr Res 19:671–682
21. Kasselouri V, Tsakiridis P, Malami C, Georgali B, Alexandridou C (1995) A study on the hydration products of a non-expansive sulphoaluminate cement. Cement Concr Res 25:1726–1736
22. Winnefeld F, Lothenbach B (2010) Hydration of calcium sulphoaluminate cements: experimental findings and thermodynamic modelling. Cement Concr Res 40:1239–1247
23. Cecille L, Kertesz C (1991) Treatment and conditioning of radioactive incinerator ashes. Elsevier Science Publishers Ltd., London
24. Arliguie G, Grandet J (1990) Study of cement hydration in presence of zinc – influence of gypsum content. Cement Concr Res 20:346–354
25. Fernandez Olmo I, Chacon E, Irabien A (2001) Influence of lead, zinc, iron (III) and chromium (III) oxides on the setting time and strength development of Portland cement. Cement Concr Res 31:1213–1219
26. Drouin M (1994) Enrobage de déchets Melox pauvres en plutonium dans une matrice de liants hydrauliques. Mémoire CNAM, Paris
27. Berger S, Cau-dit-Coumes C, Le Bescop P, Damidot D (2009) Hydration of calcium sulphoaluminate cement by a $ZnCl_2$ solution: investigation at early age. Cement Concr Res 39:1180–1187
28. Macias A, Kindness A, Glasser FP (1996) Corrosion behaviour of steel in high-alumina cement mortar cured at 5, 25 and 55-degrees-C – chemical and physical factors. J Mater Sci 31:2279–2289
29. Albino V, Cioffi R, Marroccoli M, Santoro L (1996) Potential application of ettringite generating systems for hazardous waste stabilization. J Hazard Mater 51:241–252

30. Gougar MLD, Scheetz BE, Roy DM (1996) Ettringite and C-S-H Portland cement phases for waste ion immobilization: a review. Waste Manage 16:295–303
31. Renaudin G (1998) Crystal chemistry of a double-layered hydroxide family: the AF_m phases. PhD Thesis, Henry Poincaré University, Nancy I, France
32. McCarthy GJ, Hassett DJ, Flexuraler JA (1992) Synthesis, crystal chemistry and stability of ettringite, a material with potential applications in hazardous waste immobilization. Mat Res Soc Symp Proc 245:129–140
33. Champenois JB, Cau dit Coumes C, Poulesquen A, Le Bescop P, Damidot D (2011) Conditioning highly concentrated borate solutions with calcium sulphoaluminate cement. Proceedings of NUWCEM 2011 conference, Avignon, France
34. Mesbah A, Cau-dit-Coumes C, Renaudin G, Frizon F, Leroux F (2012) Uptake of chloride and carbonate ions by calcium monosulphoaluminate hydrate, Cem. Concr. Res. 42:1157–1165
35. Ambroise J, Pera J (2004) Immobilisation of calcium sulphate in demolition waste. In: Limbachiya MC, Roberts JJ (eds) Sustainable waste management and recycling: construction demolition waste. Thomas Telford Publishing, London, pp 174–180
36. Peysson S, Pera J, Chabannet M (2005) Immobilization of heavy metals by calcium sulphoaluminate cement. Cement Concr Res 35:2261–2270
37. Berardi R, Cioffi R, Santoro L (1997) Matrix stability and leaching behaviour in ettringite-based stabilization systems doped with heavy metals. Waste Manage 17:535–540
38. Berger S, Cau-dit-Coumes C, Champenois JB, Douillard T, Le Bescop P, Aouad G, Damidot D (2011) Stabilization of $ZnCl_2$-containing wastes using calcium sulphoaluminate cement: leaching behaviour of the solidified waste form, mechanisms of zinc retention. J Hazard Mater 194:268–276
39. Process for purifying aqueous solutions polluted by nitrate ions (1992) European patent 0 494 836 B1
40. Toyohara M, Kaneko M, Mitsutska N, Fujihara H, Saito N, Murase T (2002) Contribution to understanding iodine sorption mechanism onto mixed solid alumina cement and calcium compounds. J Nucl Sci Technol 39:950–956
41. Fryda H, Vetter G, Ollitrault-Fichet R, Boch P, Capmas A (1996) Formation of chabazite in mixes of calcium aluminate cement and silica fume used for caesium immobilization. Adv Cement Res 29:29–39
42. Setiadi A, Milestone NB, Hayes M (2004) Corrosion of aluminium in composite cements. Proceedings of the 24th cement and concrete science, University of Warwick, UK
43. Zhou Q, Milestone NB, Hayes M (2006) An alternative to Portland cement for waste encapsulation – the calcium sulphoaluminate cement system. J Hazard Mater 136:120–129
44. Hayes M, Godfrey IH (2007) Development of the use of alternative cements for the treatment of intermediate level waste. Proceedings of waste management 2007 (WM'07) conference, Tucson, AZ
45. Langton CA, Stefanko DB, Serrato MG, Blankenship JK, Griffin WB, Waymer JT, Matheny D, Singh D (2011) Use of cementitious materials for SRS reactor facility in-situ decommissioning. Proceedings of waste management 2011 (WM'11) conference, Phoenix, AZ
46. Soudée E, Pera J (2000) Mechanism of setting reaction in magnesia-phosphate cements. Cement Concr Res 30:315–321
47. Wagh AS, Singh D, Jeong SY, Strain RV (1997) Ceramicrete stabilization of low-level mixed wastes, a complete story. Proceedings of the 18th annual DOE low-level radioactive waste management conference, Salt Lake City, UT
48. Wagh AS, Jeong SY, Singh D, Strain R, No H, Wescott J (1997) Stabilization of contaminated soil and wastewater with chemically bonded phosphate ceramics. Proceedings of waste management 1997 (WM'97), Tucson, AZ
49. Singh D, Barber D, Wagh A, Strain R, Tlustochowicz M (1998) Stabilization and disposal of Argonne-west low-level mixed wastes in Ceramicrete™ waste forms. Proceedings of waste management 1998 Conference (WM'98), Tucson, AZ

50. Mayberry J, Dewitt L, Darnell R, Konynenburg R, Singh D, Schumacher R, Ericksen P, Davies J, Nakaoka R (1992) Technical areas status report for low-level mixed final waste forms, DOE/MWIP-3 vol 1
51. Wagh AS, Strain R, Jeong SY, Reed D, Krause T, Singh D (1999) Stabilization of Rocky Flats Pu-contaminated ash within chemically bonded phosphate ceramics. J Nucl Mater 265:295–307
52. Wagh AS, Singh D, Patel K, Jeong S, Park J (1999) Salt waste stabilization in chemically bonded phosphate ceramics. Final report to Mixed Waste Focus Area of U.S. Dept. of Energy
53. Kay MI, Young RA, Posner AS (1964) Crystal structure of hydroxyapatite. Nature 204:1050–1052
54. Weber WJ, Ewing RC, Meldrum A (1997) On the kinetics of alpha-decay-induced amorphization in zircon and apatite containing weapons-grade plutonium or other actinides. J Nucl Mater 250:147–155
55. Chow LC (1991) Development of a self-setting calcium phosphate cement. J Ceram Soc Jpn 99:954–964
56. Valyashko VM, Korgardo LN, Khodakpovskyi IL (1968) Solubility of hydroxyapatite. Geokhimiya 1:26–36
57. Cau-dit-Coumes C, Courtois E, Guy C, Courtois S, Prené S (2003) Mise au point de ciments phosphocalciques – application au blocage de l'activité labile du combustible nucléaire irradié en situation de stockage direct. Ecole Thématique CNRS-ATILH Physique, Chimie et Mécanique des Matériaux Cimentaires, La Colle sur Loup, France
58. Van Wazer JR (1967) Inorganic polymer chemistry. J Macromol Sci A Chem 1:29–55
59. Davidovits J (1991) Geopolymers: inorganic polymeric new materials. J Therm Anal 37:1633–1656
60. Shi C, Fernandez-Jimenez A, Palomo A (2011) New cements for the 21st century: the pursuit of an alternative to Portland cement. Cement Concr Res 41:750–763
61. Barbosa VFF, MacKenzie KJD, Thaumaturgo C (2000) Synthesis and characterization of materials based on inorganic polymers of alumina and silica: sodium polysialate polymers. Int J Inorg Mater 2:309–317
62. Provis JL, Luckey GC, Van Deventer JSJ (2005) Do geopolymers actually contain nanocrystalline zeolites? A re-examination of existing results. Chem Mater 17:3075–3085
63. Duxson P, Provis JL, Lukey GC, Mallicoat SW, Kriven WM, van Deventer JSJ (2005) Understanding the relationship between geopolymer composition, microstructure and mechanical properties. Colloids Surf A Physicochem Eng Asp 269:47–58
64. Duxson P, Fernandez-Jimenez A, Provis JL, Lukey GC, Palomo A, Van Deventer JSJ (2007) Geopolymer technology: the current state of the art. J Mater Sci 42:2917–2933
65. Van Jaarveld JGS, Van Deventer JSJ, Lorenzen L (1997) The potential use of geopolymeric materials to immobilise toxic metals: Part I. Theory and applications. Miner Eng 10:659–669
66. Perera DS, Aly Z, Vance ER, Mizumo M (2005) Immobilization of Pb in geopolymer matrix. J Am Ceram Soc 88:2586–2588
67. Palomo A, Palacios M (2003) Alkali-activated cementitious materials: alternative matrices for the immobilization of hazardous wastes – part II – stabilisation of chromium and lead. Cement Concr Res 33:289–295
68. Provis JL (2009) Immobilization of toxic waste in geopolymers. In: Provis JL, Deventer JSJ (eds) Geopolymers: structure, processing, properties and industrial applications. Woodhead, Cambridge, pp 423–442
69. Shi C, Fernandez-Jimenez A (2006) Stabilization/solidification of hazardous and radioactive wastes with alkali-activated cements. J Hazard Mater 137:1656–1663
70. Fernandez-Jimenez A, Macphee D, Lachowski EE, Palomo A (2005) Immobilization of caesium in alkaline activated fly ash matrix. J Nucl Mater 346:185–193
71. Bankowski P, Zou L, Hodges R (2004) Reduction of metal leaching in brown coal fly ash using geopolymers. J Hazard Mater 114:59–67

72. Lambertin D, Frizon F, Blachère A, Bart F (2011) Corrosion of clean Mg-Zr alloys in various basic media for waste encapsulation. Proceedings of Nuwcem 2011 conference, Avignon, France
73. Criado M, Fernandez-Jimenez A, Palomo A (2010) Effect of sodium sulphate on the alkali activation of fly ash. Cement Concr Comp 32:589–594
74. Desbats-Le Chequer C, Frizon F (2011) Impact of sulphate and nitrate incorporation on potassium- and sodium-based geopolymers: geopolymerization and materials properties. J Mater Sci 46:5657–5664
75. Khalil MY, Merz E (1994) Immobilization of intermediate-level wastes in geopolymers. J Nucl Mater 211:141–148
76. Lichvar P, Rozloznik M, Sekely S (2010) Behaviour of alumino-silicate ionorganic matrix SIAL® during and after solidification of radioactive sludge and radioactive spent resins and their mixtures. Proceeding of the IAEA meeting, 18–21 Oct, Kalpakkam, India
77. Xu H, Van Deventer JSJ (2000) The geopolymerisation of alumino-silicate minerals. Int J Miner Process 59:247–266
78. Siemer DD, Olanrewaju J, Scheetz B, Grutzeck MW (2001) Development of hydroceramic waste forms for INEEL calcined waste. Ceram Trans 119:391–398

Chapter 17
Leaching of a Simulated $ZnCl_2$-Rich Radwaste Stabilized with Calcium Sulfoaluminate Cement: Experimental Investigation and First Attempt of Modeling

S. Berger, G. Aouad, C. Cau-dit-Coumes, P. Le Bescop, and D. Damidot

1 Introduction

Cementitious materials intended for radioactive waste solidification and stabilization usually include substantial amounts of ordinary Portland cement (OPC) in their formulation. However, wastes produced by nuclear activities are very diverse and some of their components may chemically react with cement phases or mixing water, thus reducing the quality of the product. For instance, ashes resulting from the incineration of technological wastes with neoprene and polyvinylchloride may contain substantial amounts of soluble zinc chloride [1]. This compound is known to have deleterious effects on OPC hydration. Setting is strongly delayed and can even be inhibited at high zinc loadings [2], while hardening is slowed down [3]. It has been shown recently that calcium sulfoaluminate (CSA) cements may exhibit a much better compatibility with zinc chloride than OPC [4, 5]: their setting is never inhibited, even at high $ZnCl_2$ (0.5 mol/L) in the mixing water, and zinc is readily insolubilized. Blending the CSA clinker with 20 % gypsum is beneficial for several reasons. (1) The hydration delay observed for a gypsum-free binder is suppressed. (2) The temperature rise and cumulative heat produced during hydration are reduced. (3) The compressive strength of the hardened materials is improved and their expansion

S. Berger (✉) • C. Cau-dit-Coumes
Commissariat à l'Energie Atomique et aux Energies Alternatives, DEN/MAR/DTCD/SPDE, Bagnols-sur-Cèze, France
e-mail: stephane.berger@kerneos.com

G. Aouad • D. Damidot
Ecole des Mines de Douai, MPE-GCE, Douai, France

Université Lille Nord de France, Lille, France

P. Le Bescop
Commissariat à l'Energie Atomique et aux Energies Alternatives, DEN/DANS/DPC/SCCME, Gif-sur-Yvette, France

F. Bart et al. (eds.), *Cement-Based Materials for Nuclear Waste Storage*,
DOI 10.1007/978-1-4614-3445-0_17, © Springer Science+Business Media New York 2013

under wet curing strongly limited. (4) Their mineralogy is less dependent on the thermal history at early age.

This chapter complements these results by investigating the behavior under leaching of a $ZnCl_2$-rich waste stabilized with CSA cement according to a test developed to understand and model the degradation processes of cement pastes [6].

2 Experimental

2.1 *Materials*

CSA cement was prepared by mixing a ground industrial CSA clinker (composition in Table 17.1; $d_{10} = 2.7$ µm, $d_{50} = 17.6$ µm, $d_{90} = 50.8$ µm, BET specific surface area = 1.3 m^2/g) with the appropriate amount of analytical grade gypsum (20 % by weight of cement; $d_{10} = 5.4$ µm, $d_{50} = 19.6$ µm, $d_{90} = 50.3$ µm, BET specific surface area = 0.4 m^2/g) for 15 min. In the clinker, ye'elimite predominated over belite and mayenite. The other minor constituents, mainly phases containing titanium and iron, could be regarded as hydraulically inactive.

The simulated waste form was prepared with a standard laboratory mixed (European standard EN 196-1) by mixing cement with a 0.5 mol/L $ZnCl_2$ solution (liquid/solid mass ratio of 0.55) at low speed for 3 min and at high speed for 2 min. The zinc concentration was representative of the concentration released in mixing water by actual radioactive incinerator ashes. The resulting paste, containing 1.2 wt % Zn, was then cast into airtight polypropylene boxes and submitted to a brief thermal excursion at early age simulating the temperature rise and fall occurring in a 200 L drum of cemented waste, as described elsewhere [4]. The thermal cycle was justified by the fact that the physicochemical evolution of CSA-based materials can be highly dependent on their thermal history at early age [7]. The samples were then cured for 3 months in sealed bags at 20°C.

2.2 *Leaching Test*

Two cylinders (5 cm in diameter, 3 cm high, protected against lateral degradation by a polymer coating) of paste specimens were leached in deionized water thermoregulated at 20 ± 1°C and kept under N_2 atmosphere to avoid carbonation.

Table 17.1 Mineralogical composition of the investigated CSA clinker (Belitex KTS 100)

Minerals	$C_4A_3\underline{S}$	C_2S	$C_{12}A_7$	CT	Periclase	$C\underline{S}$	Quartz	Others[a]
Wt%	68.5	15.9	9.5	2.9	1.5	0.5	0.5	2.4

Cement notations: C = CaO, A = Al_2O_3, S = SiO_2, $\underline{S}$ = SO_3, T = TiO_2, H = H_2O

[a]Include 1.2 % of iron oxide

The ratio between the sample surface area and the solution volume was fixed at 0.23 dm^2/L. The composition of the leaching solution remained constant during the tests, with a pH maintained at 7.0 by addition of nitric acid (0.25 mol/L). The leaching solution was renewed when the volume of added nitric acid reached 1 % of the solution volume (18 renewals were realized over the 90 d leaching period). After each renewal, the solution was analyzed by ion chromatography and OES-ICP. The mineralogy of the leached zone was assessed by progressively scraping the samples from the external surface to the sound core using a micro milling machine. With the help of XRD, TGA, and SEM/EDS, it was possible by this way to determine the phases within slices approximately 100 μm thick, parallel to the leached surface.

2.3 Characterization Methods

Crystallized phases were identified by XRD (Panalytical X'Pert Pro—Cu anode $\lambda_{K\alpha 1} = 1.54056$ Å; 40 mA and 40 kV). The acquisition range was from 5° to 60° 2θ in 0.02° 2θ steps with integration at the rate of 50 s per step. Thermogravimetric analyses were carried out under N_2 atmosphere on 50 ± 2 mg of sample using a TGA/DSC Netzsch STA 409 PC instrument at 10°C/min up to 1,000°C. The microstructure and chemical composition of the cement pastes were investigated using SEM (JEOL JSM-5910 LV, tungsten filament, or ESEM XL30-FEI, thermal field emission gun). X-ray microanalyses were performed using an energy dispersive X-ray system (EDX EDAX type). Total water porosity ϕ_w of the samples before leaching was estimated by measuring the total water amount removed from water-saturated paste samples after drying at 60°C (to limit ettringite degradation) until stable mass loss [Eq. (17.1)].

$$\phi_w(\%) = \frac{m_a - m_d}{m_a - m_w} \times 100 \tag{17.1}$$

where m_a and m_w are the water-saturated sample mass values measured in air and under water, respectively, and m_d the mass of the dried sample measured in air.

2.4 Model Features and Thermodynamic Data

The reactive transport code HYTEC [8] was used to simulate the leaching tests, taking into account diffusive transport of solutes and chemical reactions. Transport and chemistry were coupled through a sequential iterative algorithm. Moreover, the effective diffusion coefficient changed when mineral precipitation or dissolution modified the local porosity. A modified version of the Archie's law, implemented in

HYTEC, was used in a first attempt to model the retroactive effect of chemistry on mass transport.

$$D_e(\omega) = D_e(\omega_0)\left(\frac{\omega - \omega_c}{\omega_0 - \omega_c}\right)^{\alpha} \tag{17.2}$$

where ω = porosity, D_e = effective diffusion coefficient, ω_c = critical porosity threshold under which diffusion stops, α = empirical Archie coefficient, ω_c and α were fixed to 0.005 and 2, respectively.

1D grid was used with a 100 μm mesh large. Zero-flux boundary conditions were defined. The cement paste area had a section of 0.393 dm^2. The leachant chemistry corresponded to a pure water solution at pH 7 and 20°C, in agreement with the experimental conditions. The leachant concentrations were maintained at zero. The degradation was modeled over 90 days. The solubility constants (log K) for CSA cement hydrates were previously given in [7], those of the Kuzel's salt and Friedel's salt were, respectively, fixed to −27.11 and −26.31 at 25°C [5]. Data relative to zinc-containing phases were found in [9] and [10].

3 Results and Discussion

3.1 *Characterization of the Samples Before Leaching*

The hydrate assemblage observed on the 90-day-old samples (Fig. 17.1) comprised ettringite, amorphous aluminum hydroxide, and Friedel's salt, an AFm-structured compound in which the positively charged main layers are balanced by the insertion of chloride anions in the interlayer ($C_3A{\cdot}CaCl_2{\cdot}10H_2O$). Small amounts of residual ye'elimite and gypsum were noticeable. No zinc-containing crystallized phases could be detected. The total water porosity was measured to be 32 %.

3.2 *Durability Under Leaching by Deionized Water*

3.2.1 Characterization of the Leachates

The cumulative quantities of OH^-, Ca^{2+}, SO_4^{2-}, and Cl^- released in the leaching solution increased linearly as a function of the square root of time (Fig. 17.2). Leaching was controlled by diffusion. Silicates and aluminates were not detected in the solution. However, an amorphous white compound, identified as aluminum hydroxide by EDX microanalysis, precipitated in the leachate. The zinc concentration always remained below the detection limit of the ICP method (2 μmol/L), meaning that, for the whole test (18 renewals), the leached fraction of zinc was <0.1 %. The cement matrix thus provided good confinement of zinc.

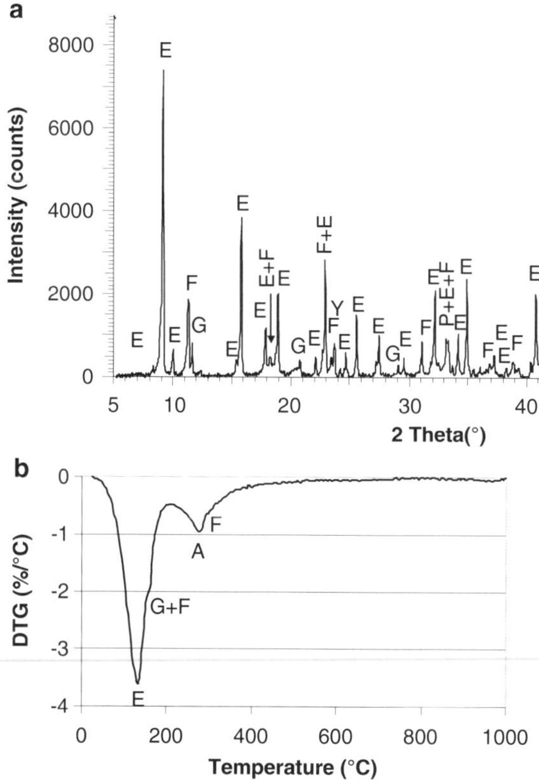

Fig. 17.1 Mineralogy before leaching of 90-day-old samples investigated by (**a**) X-ray diffraction, and (**b**) thermogravimetry (*E* ettringite, *F* Friedel's salt, *G* gypsum, *Y* ye'elimite, *P* perovskite, *A* aluminum hydroxide)

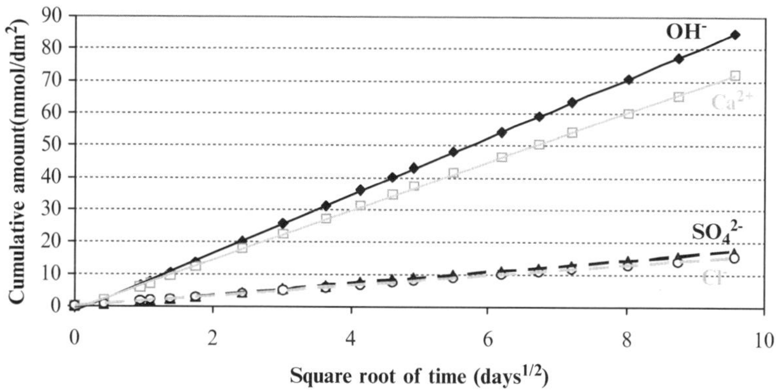

Fig. 17.2 Cumulative amounts of Ca^{2+}, SO_4^{2-}, Cl^-, and OH^- released during leaching of the sample in pure deionized water (pH 7, 20 °C). The corresponding fluxes were, respectively, 9.03 ± 0.04 mmol/dm^2/day$^{0.5}$ (OH^-), 7.63 ± 0.03 mmol/dm^2/day$^{0.5}$ (Ca^{2+}), 1.82 ± 0.01 mmol/dm^2/day$^{0.5}$ (SO_4^{2-}), and 1.59 ± 0.01 mmol/dm^2/day$^{0.5}$ (Cl^-)

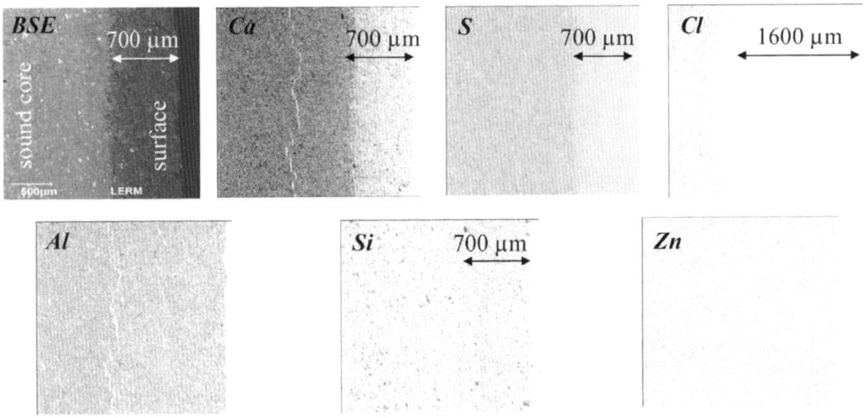

Fig. 17.3 SEM characterization of the leached cement pastes after 3 months of leaching (deionized water, pH 7, 20 °C)

3.2.2 Characterization of the Degraded Solid

The position of the degradation front in the samples was assessed after 3 months of leaching. Portlandite is usually a good indicator for the location of the degradation front for common OPC-based materials [6]. However, the tracer for the degradation of CSA-based materials was unknown. Different techniques were thus combined: SEM observations and X-ray microanalysis, as well as X-ray diffraction.

The first approach was to use chemical contrast from SEM/BSE images, the density of the degraded zone being lower than that of the sound core due to decalcification. A sharp transition was effectively observed between a bright zone (sound core) and a dark one (degraded material) (Fig. 17.3). The degradation depth was around 700 µm. Ca- and S-mapping clearly showed the decalcification and sulfur loss from the cement paste near the surface exposed to leaching, in a zone with a thickness of 700 µm, which was in good agreement with the estimation derived from the BSE image. The silicate density in this zone appeared to increase slightly. The Cl mapping revealed that chlorides were leached within a larger domain, to a depth of 1,600 µm from the surface. The alumina and zinc contents appeared to remain relatively constant whatever the considered depth.

In a second approach, XRD analyses were carried out on the samples surface which was scraped off step by step (thickness ≈ 100 µm for each step) to obtain XRD profiles (Fig. 17.4). In addition to ettringite and Friedel's salt detected in the cement pastes before leaching, the sound core contained Kuzel's salt, an AFm phase with ordered chloride and sulfate anions in its interlayer ($C_3A \cdot ½CaSO_4 \cdot ½CaCl_2 \cdot 11H_2O$), and strätlingite ($C_2ASH_8$). These new hydrates resulted from a restart of hydration of the residual anhydrous phases after immersion of the samples previously cured in sealed bag.

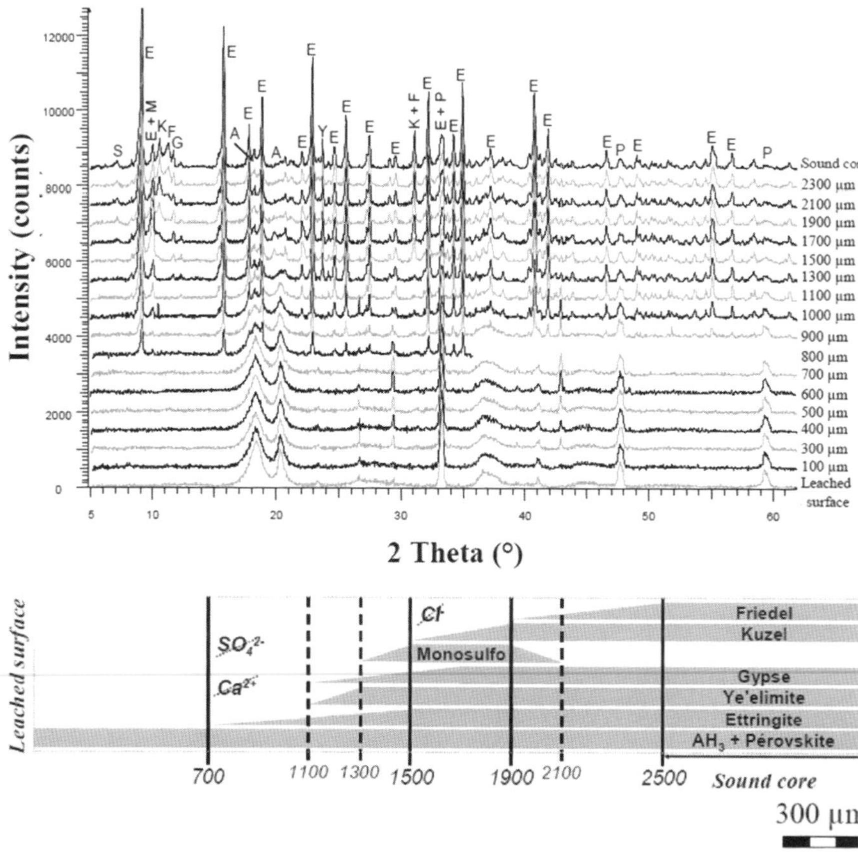

Fig. 17.4 XRD patterns of the degraded zone of the samples after 3 months of leaching (*S* strätlingite, *E* ettringite, *M* monosulfoaluminate, *K* Kuzel's salt, *F* Friedel's salt, *G* gypsum, *A* aluminum hydroxide, *Y* ye'elimite, *P* perovskite)

The first signs of alteration were noticed from a depth of 2,500 μm. Several processes were observed:

- Dissolution of Friedel's salt (from 2,500 to 1,900 μm) and Kuzel's salt (from 1,900 to 1,500 μm) which released chlorides
- Transient precipitation of monosulfoaluminate (from 2,100 to 1,300 μm)
- Dissolution of monosulfoaluminate (from 1,500 to 1,300 μm) and of residual ye'elimite (from 1,300 to 1,100 μm)
- Dissolution of ettringite (from 1,500 to 700 μm)

The surface layer, with a depth of 700 μm (consistent with the SEM observations), was mainly composed of perovskite, poorly crystallized aluminum hydroxide, and probably C-A-S-H. The latter were not identified with certainty, but

Table 17.2 Mineralogical composition used for modeling

Minerals	Ettringite	AH_3 gibbsite	Friedel's salt	Kuzel's salt	Inert
Wt%	53.8	20.4	7.6	5.2	13.0

the leached silicate flux was below the detection limit and STEM analyses performed on the surface layer showed the presence of small amounts of calcium and silicon, together with aluminum. In addition, TGA curves exhibited a weight loss at 98°C which could correspond to C-A-S-H. This surface layer was highly porous and friable.

Zinc was homogeneously distributed in all the three zones, which rules out the assumption of dissolution in the intermediate zone, followed by reprecipitation in the surface layer. Additional characterization using TEM and STEM in SEM of the hydrates in the different zones [5] showed that the confinement of zinc by the cement matrix likely resulted from chemisorption of Zn^{2+} onto aluminum hydroxide (which involved the binding of Zn^{2+} to AlO_6 in an edge-sharing bidentate geometry), rather than $Zn^{2+} \leftrightarrow Ca^{2+}$ substitution in the structure of ettringite or AFm phases, as previously postulated. Sorption of Zn^{2+} onto C-(A)-S-H could also occur, but could not explain by itself the good retention of zinc.

3.3 *Modeling: Sample Composition and Results*

The composition of the 90-day-old hydrated cement pastes before leaching was simplified for modeling (Table 17.2). The residual anhydrous phases were considered as inert during leaching (e.g., C_2S and ye'elimite). In addition, the small amount of strätlingite observed experimentally in the sound core was neglected. The porosity of the sample corresponded to its experimental value: 32 %. The effective diffusion coefficient was adjusted to 6.5×10^{-11} m/s to reproduce the cumulative amount of Cl^- released during leaching (15.5 mmol/dm). The main objective of modeling was to predict the mineralogical evolutions during leaching. Results are given in Fig. 17.5 and Table 17.3.

The degradation front, which was determined by the total depletion of Friedel's salt, occurred at a depth of 4,800 $\pm$ 100 μm. This is not in agreement with experiment (measured depth of 1,900 μm). Similarly, the degradation front of Kuzel's salt was predicted to occur at 4,000 $\pm$ 100 μm and was experimentally observed at 1,500 μm. Modeling the chloroaluminate phases degradation has still to be improved.

However, the model was closer to experiment for calcium monosulphoaluminate hydrate and ettringite, which were fully depleted at 1,600 $\pm$ 100 μm (1,300 μm experimentally) and at 800 $\pm$ 100 μm (700 μm experimentally), respectively. The mineralogical evolutions during leaching were also well predicted, which was the main objective of the modeling. In particular, the transient precipitation of monosulphoaluminate was reproduced.

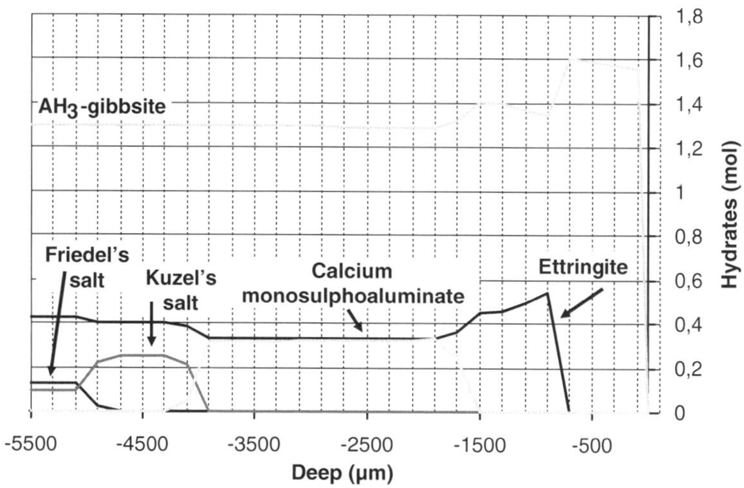

Fig. 17.5 Modeling of phase assemblage evolution of a 90-day-old cement paste during leaching by pure water (pH 7, 20 °C) using the reactive transport code HYTEC

Table 17.3 Comparison between the cumulative amounts of Ca^{2+}, SO_4^{2-}, Cl^-, and OH^- released during leaching and those obtained by modeling

Cumulative amount	Ca^{2+}	SO_4^{2-}	Cl^-	OH^-
Experiment ($mmol/dm^2$)	72.0	17.0	15.5	84.4
Modeling ($mmol/dm^2$)	36.3	9.5	15.2	28.2

The effective diffusion coefficient was adjusted to reproduce the cumulative amount of Cl^- released, which explains the good agreement between experiment and modeling. However, the others fluxes values were largely underestimated. This could result from several factors: (1) dissolution of the residual anhydrous phases was not taken into account by the model (in particular, C_2S was considered as inert), (2) precipitation of AH_3 in the leaching solution was not reproduced, and (3) the likely precipitation of C-(A)-S-H was omitted.

4 Conclusion

The simulated cemented waste form (cement paste prepared with a CSA binder comprising 20 % gypsum and a 0.5 mol/L $ZnCl_2$ mixing solution) exhibited three zones after 90 days of leaching by deionized water at fixed pH and temperature (7 and 20°C, respectively):

- The surface layer, with very poor mechanical strength, composed of aluminum hydroxide, perovskite, and C-A-S-H, which could be easily detected from BSE

images, Ca- and S-mapping (EDS analysis), and X-ray diffraction (disappearance of the ettringite signal)
- A less porous intermediate zone in which several precipitation and dissolution fronts occurred, as shown by X-Ray diffraction and Cl mapping (EDS analysis)
- The sound core

The first attempt of modeling the leaching test using the reactive transport code HYTEC was promising since the phase assemblage evolution was well reproduced. However, the extent of degradation was largely overestimated, and the model still needs to be improved.

References

1. Cecille L, Kertesz C (eds) (1991) Treatment and conditioning of radioactive incinerator ashes. Elsevier Science Publishers Ltd., London, 239 p
2. Arliguie G, Grandet J (1990) Study of cement hydration in presence of zinc – influence of gypsum content. Cement Concr Res 20:346–354
3. Fernandez OI, Chacon E, Irabien A (2001) Influence of lead, zinc, iron (III) and chromium (III) oxides on the setting time and strength development of Portland cement. Cement Concr Res 31:1213–1219
4. Berger S, Cau-dit-Coumes C, Le Bescop P, Damidot D (2009) Hydration of calcium sulfoaluminate cement by a $ZnCl_2$ solution: investigation at early age. Cement Concr Res 30:1180–1187
5. Berger S (2009) Etude des potentialités des ciments sulfo-alumineux bélitiques pour le conditionnement du zinc: de l'hydratation à la durabilité, PhD thesis, Lille 1 University, 248 p
6. Adenot F, Buil M (1992) Modelling of the corrosion of cement paste by deionized water. Cement Concr Res 22:489–495
7. Berger S, Cau-dit-Coumes C, Le Bescop P, Damidot D (2011) Influence of a thermal cycle at early age on the hydration of calcium sulphoaluminate cements with variable gypsum contents. Cement Concr Res 41:149–160
8. Van der Lee J, De Windt L, Lagneau V, Goblet P (2003) Module-oriented modeling of reactive transport with HYTEC. Comput Geosci 39:265–275
9. Allison JD, Brown DS, Novo-Gradac KJ (1991) MINTEQA2/PRODEF2, A geochemical assessment model for environmental systems: version 3.0 user's manual. EPA/600/3-91/021 Athens, GA, 30605: EPA (USA)
10. Ziegler F, Johnson CA (2001) The solubility of calcium zincate ($CaZn_2(OH)_6.2H_2O$). Cement Concr Res 31:1327–1332

Chapter 18
Conditioning Highly Concentrated Borate Solutions with Calcium Sulfoaluminate Cement

J.B. Champenois, C. Cau-dit-Coumes, A. Poulesquen, P. Le Bescop, and D. Damidot

1 Introduction

Calcium silicate cements, such as Ordinary Portland cement (OPC), are widely used for low- and intermediate-level radioactive waste conditioning. However, the wastes produced by nuclear industry are very diverse and some of their components may chemically react with cement phases. Among these components, soluble borates are well known to interfere with hydration of OPC: they act as set retarders and false set may also occur [1]. Two main strategies have been used to solidify wastes with high contents of borates:

- Addition of calcium hydroxide to OPC to precipitate borates as calcium hexahydroborite $CaO.B_2O_3.6H_2O$ [2–4].
- Addition of calcium hydroxide and aluminates to OPC to incorporate borates into the AFt phase $3CaO.Al_2O_3.2Ca[B(OH)_4]_2.Ca(OH)_2.30H_2O$ (calcium quadriboroaluminate QBA or high-boron-content AFt) [5, 6].

The former has been used to encapsulate PWR evaporator concentrates with boron up to 40 g L^{-1}. However, two disadvantages arise: (1) cement set is still retarded and (2) calcium hexahydroborite is unstable in cement and is progressively

J.B. Champenois • C. Cau-dit-Coumes (✉) • A. Poulesquen
Commissariat à l'Energie Atomique et aux Energies Alternatives, DEN/MAR/DTCD/SPDE, BP17171, 30207 Bagnols-sur-Cèze, France
e-mail: celine.cau-dit-coumes@cea.fr

P. Le Bescop
Commissariat à l'Energie Atomique et aux Energies Alternatives, CEA/DEN/SAC/DPC/SCCME, 91192 Gif/Yvette, France

D. Damidot
Ecole des Mines de Douai, MPE-GCE, Douai, France

Université Lille Nord de France, 59000 Lille, France

F. Bart et al. (eds.), *Cement-Based Materials for Nuclear Waste Storage*,
DOI 10.1007/978-1-4614-3445-0_18, © Springer Science+Business Media New York 2013

converted into calcium mono- or quadri-boroaluminate. The latter, which is more effective to reduce setting inhibition, was applied to the conditioning of ion exchange resins and highly concentrated borate solutions.

An alternative may consist in using a binder showing an improved chemical compatibility with the waste, such as calcium sulfoaluminate cement (CSA). The expected beneficial effects would be a simplification of the solidification process and a possible increase in the waste incorporation rate. When hydrating, CSA cements form significant amounts of ettringite and calcium monosulfoaluminate hydrate depending on their gypsum content. It is well known that ettringite can incorporate borate ions in its structure [7, 8]. Moreover, an AFm phase containing borate anions in its interlayer has already been described [9]. In this work, we first studied the influence of borate anions in the mixing solution on the early age hydration of three CSA cements comprising, respectively, 0, 10 and 20 % of gypsum. Evolution of the mineralogy of the CSA cement pastes was then characterized over 90 days. Finally, the length change and mechanical properties of mortars comprising 0, 10 and 20 % of gypsum were monitored over 180 days.

2 Experimental Section

2.1 Materials and Specimen Preparation

Three CSA cements were prepared by mixing a ground industrial CSA clinker (d_{50} = 15.1 μm, specific surface area = 4.8 m^2/g) with the appropriate amount (0 %, 10 % and 20 % by weight of cement) of analytical grade gypsum (d_{50} = 18.7 μm, specific surface area = 0.5 m^2/g) for 15 min. Mineralogical composition of the CSA clinker is reported in Table 18.1. In the clinker, ye'elimite predominated over belite and mayenite. The other minor constituents, mainly phases containing titanium and iron, could be regarded as hydraulically inactive.

Cement pastes and mortars were prepared using a *w/c* ratio of 0.6. A blend of two siliceous sands (0.1–1.2 mm) with a sand-to-cement ratio of 3 was used to optimize the workability and limit the heat release of fresh mortars during hydration. The mixing solution was prepared by dissolving the appropriate amount of analytical grade boric acid in distilled water (1 mol/L of boric acid); pH of the mixing solution was then raised up to 11 by adding analytical grade sodium hydroxide (960 μmol/L of sodium hydroxide). Mixing was performed in a standardized laboratory mixer (following European standard EN 196-1) at low speed for 3 min and at high speed for 2 min. Cement pastes were cast into airtight polypropylene boxes (7 mL of paste per box) and cured at 20°C for three months. Mortars were cast into 4 × 4 × 16 cm moulds and cured for 7 days at 20 ± 1°C and 95 ± 5 % R.H. The specimens were then demoulded, measured and kept at room temperature in sealed bags or immersed under water at 20 ± 1°C.

Table 18.1 CSA clinker mineralogical composition (Belitex KTS 100)

Minerals	$C_4A_3\check{S}$	C_2S	$C_{12}A_7$	CT	MgO	Others
Wt %	66.2	13.5	9.4	2.6	0.9	7.4

Table 18.2 Shorthand notations of the investigated compositions

Gypsum content (% by wt of cement)	$[B(OH)_3]_{total}$ (mol/L)	Paste	Mortar
0	0	C 0G	M 0G
	1	C 0G 1M	M 0G 1M
10	0	C 10G	M 10G
	1	C 10G 1M	M 10G 1M
20	0	C 20G	M 20G
	1	C 20G 1M	M 20G 1M

Shorthand notations were used to refer to the different investigated compositions, as summarized in Table 18.2.

2.2 *Characterization of Early Age Hydration*

Hydration of pastes was followed using an isothermal microcalorimetry at 25°C (SETARAM, C80 type) and a AR1000 Rheometer using a ribbon geometry in dynamic mode (strain of 10^{-4} and stress of 1 rad/s). Hydration was stopped after fixed periods of time (from 5 min to 3 months) by successively immersing the crushed pastes into isopropanol and drying them in a controlled humidity chamber (with 20 % relative humidity at 22 ± 2°C). Crystallized phases were identified by X-ray diffraction (Siemens D8—copper anode $\lambda_{K\alpha 1} = 1.54056$ Å generated at 40 mA and 40 kV) on pastes ground to a particle size of less than 100 μm. The acquisition range was from 5° to 60° 2θ in 0.02° 2θ steps with integration at the rate of 50 s per step. Thermogravimetric analyses were carried out under N_2 atmosphere on 50 ± 2 mg of sample using a TGA/DSC Netzsch STA 409 PC instrument at 10°C/min up to 1,000°C. Pore solution was extracted as far as possible using a 2-ton press in order to measure its pH.

2.3 *Characterization of Mortar Properties*

The compressive strength of mortar prisms cured in sealed bag or under water was measured following European standard EN 196-1 after 1 day, 7 days, 28 days, 90 days and 180 days. Two specimens were used at each testing age. Dimensional variations were measured after 1 day, 7 days, 28 days, 90 days and 180 days using a retractometer.

3 Results

3.1 Investigation of Early Age Hydration: Influence of the Gypsum Content of the Binder and of the pH of the Mixing Solution

The heat flux released during hydration of the cement pastes by a 1 mol/L $B(OH)_3$ solution adjusted at pH = 11 was first investigated as a function of the gypsum content (Fig. 18.1) using isothermal calorimetry at 25°C. Increasing the gypsum content resulted in a longer induction period. When the mixing solution contained 1 mol/L of borates at pH = 11, the heat flux recorded by isothermal calorimetry reached its maximum after 28 h only for a gypsum-free binder, instead of more than 80 h for a binder comprising 10 % of gypsum and about 95 h for a binder comprising 20 % of gypsum. Under similar conditions, the maximum heat flux would occur after about 140 h for OPC. A retarding effect by borate anions was still observed when using CSA cements since the setting time of CSA cements mixed with pure water under similar conditions is always less than 20 h.

From a mineralogical point of view, for both C 0G 1M and C 10G 1M compositions, different events occurred before the massive dissolution of anhydrous phases, corresponding to the main peak of heat flux. Small amounts of anhydrous phases began to dissolve, as indicated by the less intense peak at about 5 h. XRD analysis showed that the dissolution of anhydrous phases was accompanied by the precipitation of small amounts of an AFt-type phase (Fig. 18.2, *top*). The peak shift of the {001} reflection could result from the incorporation of borate anions in the AFt phase. In addition to this crystallized

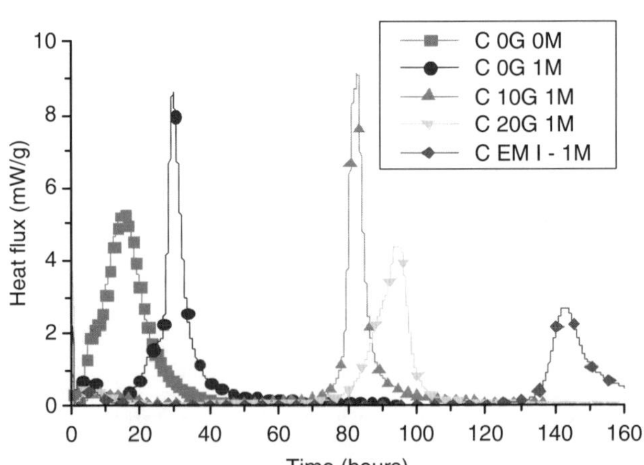

Fig. 18.1 Influence of the initial gypsum content on the heat flow generated during the hydration of CSA cements by a boric acid solution ([B] = 1 mol/L, pH = 11—*w*/*c* = 0.6)

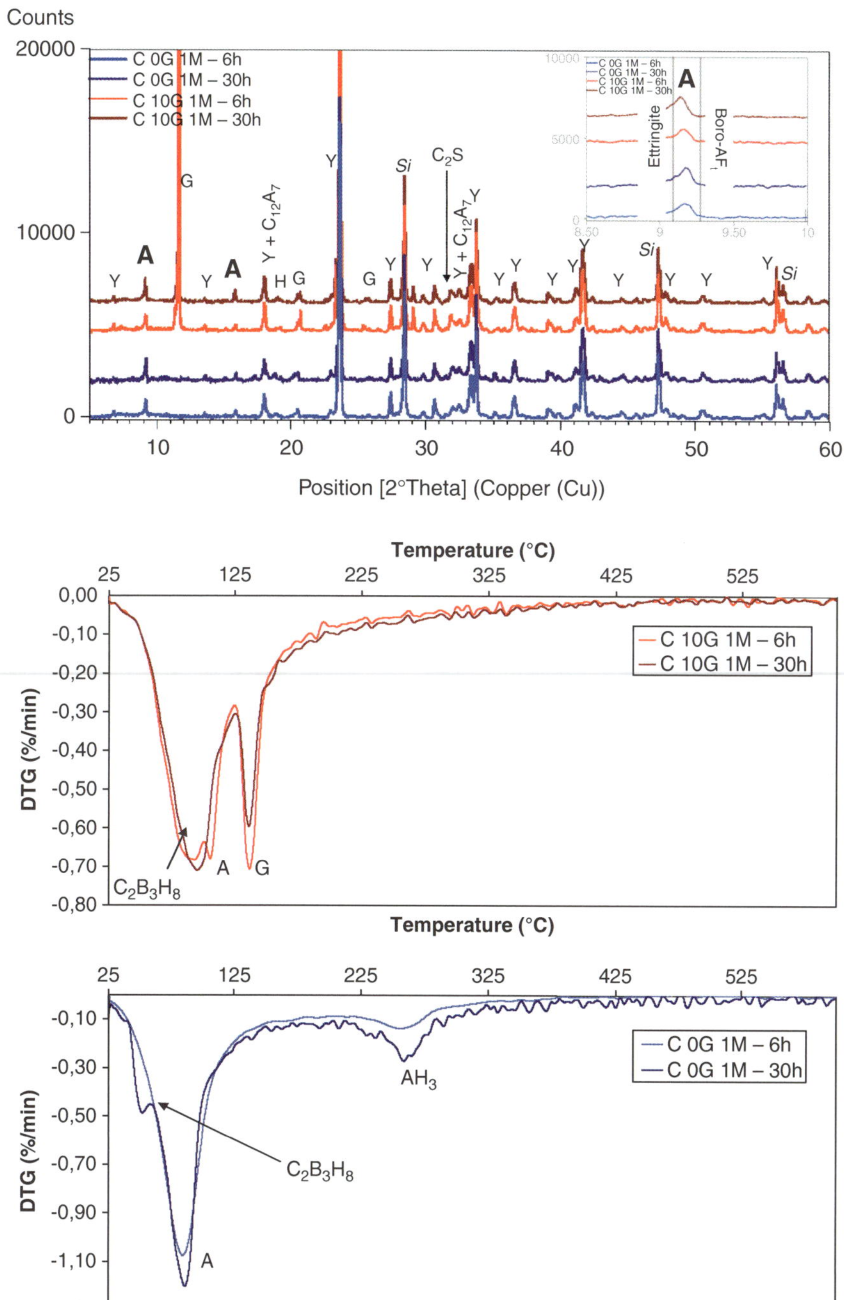

Fig. 18.2 Mineralogical compositions of C 0G 1M and C 10G 1M pastes at 6 h and 30 h after the beginning of hydration (*Top*: XRD patterns; *Bottom*: DTG analysis; *A* AFt phases, *Y* ye'elimite, *G* gypsum, *H* AH_3)

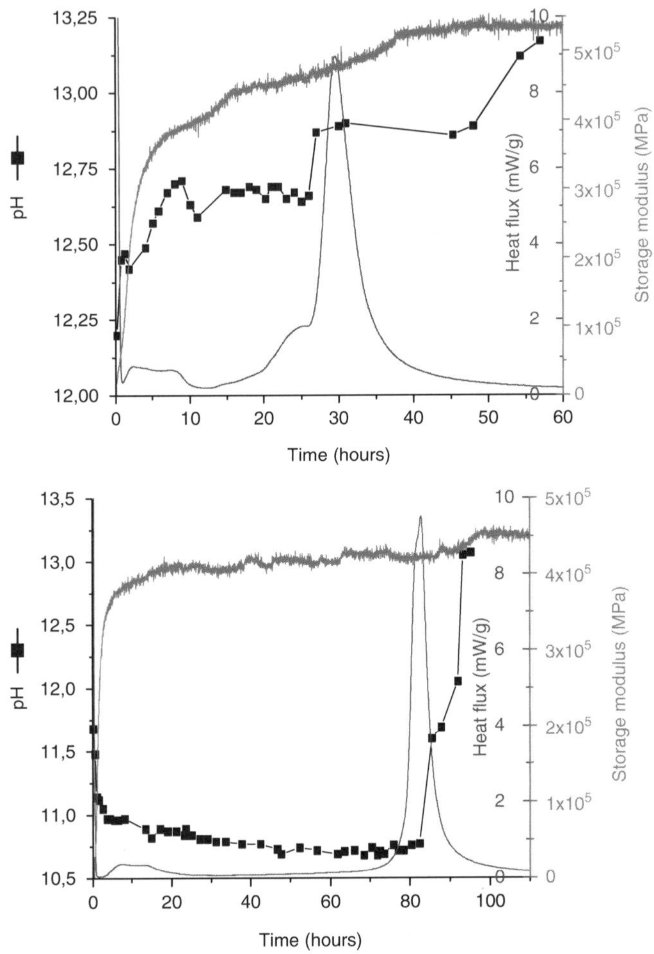

Fig. 18.3 Comparison between pore solution pH evolution, heat flux and storage modulus evolution during hydration for C 0G 1M (*top*) and C 10G 1M (*bottom*)

phase, TGA analysis showed the likely precipitation of an amorphous compound characterized by a weight loss at about 70°C (Fig. 18.2, *bottom*).

The pore solution of the investigated pastes was extracted before the massive dissolution of the anhydrous phases in order to obtain a mean value of its pH. The mean pH values were 12.6 ± 0.2 for C 0G 1M (compared to 11.4 ± 0.1 for reference C 0G), instead 10.8 ± 0.2 for a cement containing 10 % of gypsum (compared to 10.2 ± 0.2 for C 10G).

These results make us to consider the precipitation of the amorphous calcium borate compound $C_2B_3H_8$ to explain the retarding effect of borate anions on CSA cements hydration, as many authors did to explain the retarding effect of borate anions on OPC and calcium aluminate cements [10, 11].

- *C 10G 1M composition*. The pH of the pore solution during the induction period (close to 11) is consistent with the precipitation of the amorphous compound, which was supported by TG analysis (Fig. 18.2). The rise of the pore solution pH occurs simultaneously with the massive dissolution of the anhydrous phases, causing the destabilization of the amorphous compound and the precipitation of a borate-containing AFt phase (Fig. 18.2). Additional results obtained by dynamic mode rheometry (Fig. 18.3, *bottom*) showed that the precipitation of the amorphous compound occurs at very early age and gives cohesion to the paste. This compound would be responsible for the false set observed during hydration of CSA cements by borate solutions. The subsequent precipitation of hydrated phases, consecutive to the amorphous compound destabilization and to the massive dissolution of anhydrous phases, yields to a strength gain, which is shown by the increase in the storage modulus value in Fig. 18.3.
- *C 0G 1M composition*. According to the results of Casabonne, $C_2B_3H_8$ would not be stable during the induction period because of the too high pH of the pore solution (comprised between 12.0 and 12.8). Nevertheless, some of our observations seem to indicate the precipitation of an amorphous compound even under these conditions: (1) occurrence of a significant weight loss at 70°C by TGA carried out on a 20 h-old paste, which could not be ascribed to a crystallized phase, (2) rapid increase in the storage modulus during the first minutes after mixing, as shown by dynamic mode rheometry. Additional experiments are under way to characterize more accurately the stability domain of amorphous calcium borates. The precipitation of small amounts of such a compound would slow down the hydration, but to a lower extent than for C 10G 1M sample. Indeed, the heat flux during the induction period, was always different from zero, which indicated the continuous dissolution of anhydrous phases and the precipitation of an AFt phase, as shown by the area of the {001} XRD reflection peak which increased with time (Fig. 18.2, *top*). It also explained the constant increase in the storage modulus (Fig. 18.3, *top*), which mechanically corresponded to a strength gain.

3.2 Mineralogical Evolution over 90 days

Mixing a CSA cement free from gypsum with pure water caused the precipitation (after 7 days only) of four main hydrates: calcium monosulfoaluminate hydrate, strätlingite, aluminium hydroxide AH_3 and small amounts of ettringite. Most of the ettringite that precipitated in the first hours after mixing was converted to calcium monosulfoaluminate hydrate, as the sulphate concentration was too low to stabilize ettringite. The precipitation of strätlingite resulted from the dissolution of C_2S. Adding 10 % gypsum to the cement produced a mineralogical assemblage containing mostly ettringite, small amounts of calcium monosulfoaluminate hydrate, aluminium hydroxide and strätlingite. The higher concentration of sulphates stabilized the AFt phase versus the AFm phase.

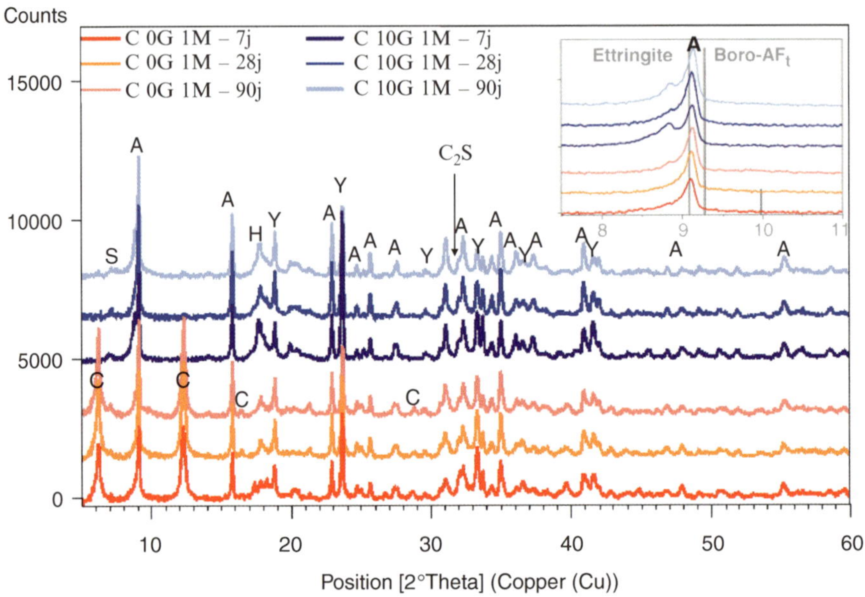

Fig. 18.4 Mineralogical compositions of C 0G 1M and C 10G 1M pastes at 7 days, 28 days and 90 days after the beginning of hydration (Y = ye'elimite, A = AFt phases, H = AH_3, S = strätlingite, C = CAH_{10})

Mineralogical assemblages obtained when mixing CSA cements with borate solutions (1 mol/L at pH = 11) were quite different (Fig. 18.4).

Hydration of C 0G 1M produced an AFt-type phase containing borate anions, the transient hydrated calcium aluminate phase CAH_{10} and aluminium hydroxide AH_3. No AFm-type phase was observed over the 90 days of the study. Borate anions seem to have the same effect as sulphate anions: since borate anions can easily form an AFt-type phase, these anions stabilize the AFt phase as sulphate anions do with ettringite. At 28 days, only traces of strätlingite were detected. Its amount increased however at 90 days. The strong retarding effect of borates on the hydration of calcium silicate phases is well documented. Besides, the delayed occurrence of this phase shows that borate anions have a retarding effect not only on the early hydration, but also on the long-term hydration of gypsum-free CSA cement. The mineralogical evolution of this paste will have to be studied on a longer period of time to follow the possible conversion of CAH_{10} and its possible consequences on the properties of the material.

Adding 10 % gypsum to the cement seemed to yield a more stable mineralogical assembly. After 7 days, the CSA cement paste hydrated with a borate solution (1 mol/L at pH = 11) comprised aluminium hydroxide AH_3 and two AFt phases. Although borate and sulphate ions are supposed to form a complete solid solution among AFt phase [8], two distinct AFt phases were observed, each of them containing borate and sulphate ions in different ratios. Small amounts of strätlingite were already observed at this time. Increasing the hydration degree (28 and 90 days) increased the amounts of hydrates precipitated, without any change in the phase assemblage. No AFm-type phase was observed during this period.

For both C 0G 1M and C 10G 1M compositions, large amounts of anhydrous phases were still present 90 days after the beginning of hydration.

3.3 Evolution of the Macroscopic Properties

The compressive strength and length change of mortar bars cured in sealed bag or under water at ambient temperature were characterized over 180 days (Figs. 18.5 and 18.6).

Gypsum free mortars. For M 0G 1M, mortars cured in sealed bags exhibited a compressive strength of about 50 MPa at 7 days after mixing, slightly higher than that of a reference prepared with pure water (40 MPa for M 0G). Increasing the hydration degree caused a decrease in the mechanical strength of sample M 0G 1M from 50 MPa at 7 days to 40 MPa at 180 days. This decrease may have resulted from

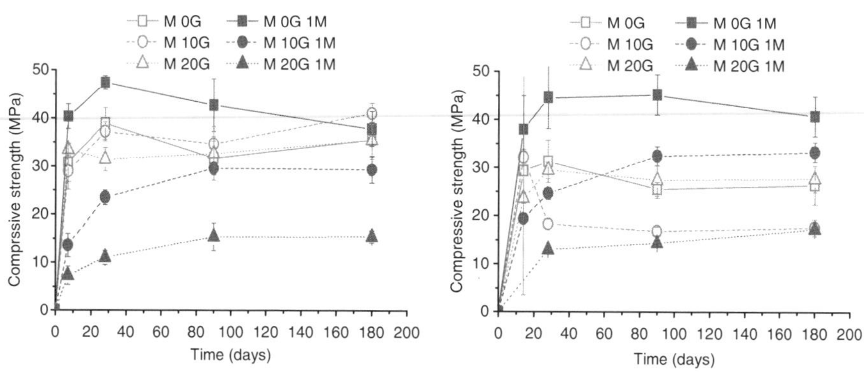

Fig. 18.5 Evolution of the compressive strength of the mortars depending on their composition and curing conditions (*left*: sealed bag, *right*: immersion under water)

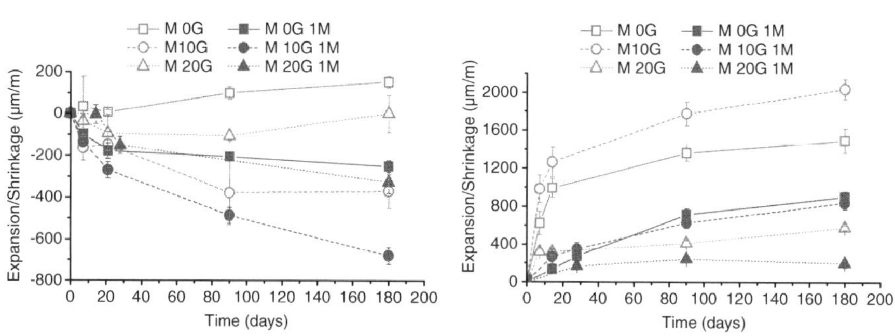

Fig. 18.6 Evolution of length change of investigated mortars depending on their composition and curing conditions (*left*: sealed bag, *right*: immersion under water)

mineralogical evolutions such as conversion of the transitory hydrate CAH_{10} and precipitation of strätlingite from C_2S and AH_3 (Fig. 18.4). The same trend was observed for mortars cured under water. From a dimensional point of view, all the borate-containing mortars cured in sealed bags exhibited shrinkage, even the gypsum-free material (−200 μm/m for M 0G 1M at 180 days). For comparison, reference M 0G slightly expanded under the same conditions (+200 μm/m for M 0G). The behaviour of sample M 0G 1M resembled that of M 10G, which may be explained by the fact that borate anions acted as sulphate anions during the hydration (the amount of gypsum in M 10G would release about 1 mol/L of sulphate ions SO_4^{2-} in the pore solution). Under water, the expansion of mortar M 0G 1M was lower than that of reference M 0G (+800 μm/m for M 1M 0G at 180 days compared to +1,200 μm/m for M 0G). Equilibrium was not reached at 180 days.
Gypsum-containing mortars (10 % and 20 %). Adding borates to a mortar prepared with a binder containing 10 % gypsum weakened the mechanical strength, but this parameter presented a monotonous evolution with time. The compressive strength of mortar M 10G 1M cured in sealed bags went on increasing from 15 MPa at 7 days to 30 MPa at 180 days, compared to 35 MPa at 7 days and 40 MPa at 180 days for reference M 10G. Curing M 10G 1M under water did not change the strength evolution. At 180 days, the compressive strength of mortars M 10G 1M seemed to be stabilized. From a dimensional point of view, the shrinkage of the borate-containing mortars cured in sealed bags was comparable to that of the reference (−600 μm/m for M 10G 1M at 180 days compared to −400 μm/m for M 10G). Curing borate-containing mortars under water resulted in a smaller expansion than for the reference (+800 μm/m for M 10G 1M at 180 days compared to +2,000 μm/m for M 10G).

Borate-containing mortars prepared with a binder comprising 20 % exhibited poor compressive strength whatever the curing conditions (10 MPa for M 20G 1M at 180 days compared to 35 MPa for M 20G). Nevertheless, shrinkage under sealed bag (about 0 μm/m for M 20G 1M at 180 days compared to 0 μm/m for M 20G) and expansion under water (+200 μm/m for M 20G 1M at 180 days compared to +400 μm/m for M 20G) were very limited and equilibrium seemed to be reached at 180 days.

4 Conclusion

From a kinetics point of view, CSA cements with low gypsum contents should be preferred to solidify aqueous solutions with high borate concentrations (1 mol/L). Adding gypsum buffers the pore solution pH at ≈10 (instead of ≈12 for a gypsum-free material) at early age, and thus favours the precipitation of amorphous calcium borate $C_2B_3H_8$. This compound significantly retards hydration by slowing down ionic exchanges necessary for anhydrous phases dissolution and hydrated phases precipitation.

Nonetheless, the presence of borates in a gypsum-free material slows down the subsequent steps of hydration since strätlingite is observed only after 28 days.

Moreover, the precipitation of the transient hydrate CAH_{10} is promoted under these conditions and large amounts of this phase are still present at 90 days. The precipitation of strätlingite and the possible conversion of CAH_{10} occurring lately may induce a decrease in the mechanical strength which may be undesirable for aged mortars. Adding high amounts of gypsum (typically 20 %) to the hydraulic binder should be avoided since the induction period is largely increased and mortars prepared with this cement develop poor mechanical strength.

Using a binder comprising 10 % of gypsum seems to be a good compromise when the mixing solution contains borate anions (1 mol/L). Even if the induction period is longer than for a gypsum-free binder, hydration remains more rapid than for OPC, and the mineralogical assemblage is more stable since CAH_{10} is almost absent 90 days after mixing. The mechanical strength continuously increases with the hydration degree. Length changes under wet-curing and sealed bag remain moderate and seem to be stabilized after 180 days.

Future studies will be focussed on the long-term evolution of these materials and on their behaviour under leaching.

Acknowledgments L. Petit, from Electricité de France, is deeply acknowledged for his support of the project.

References

1. Ramachandran VS, Lowery MS (1992) Conduction calorimetric investigation of the effect of retarders on the hydration of Portland cement. Thermochim Acta 195:373–387
2. Benavides E (1997) Immobilization of evaporator concentrates with high boron content in cement matrix. In: International symposium on nuclear energy & radioactive waste management, Bucarest, vol 2, pp 470–471
3. Hernandez S, Guerrero A, Goni S (2000) Leaching of borate waste cement matrices: pore solution and solid phase characterization. Adv Cement Res 12:1–8
4. Guerrero A, Goni S (2002) Efficiency of a blast furnace slag for immobilizing simulated borate radioactive liquid waste. Waste Manage 22:281–288
5. Le Bescop P, Bouniol P, Jorda M (1990) Immobilization in cement of ion exchange resins. In: Material Research Society symposium proceedings, vol 176, pp 183–189
6. Goni S, Guerrero A (2001) Stability of calcium aluminate cement matrices mixed with borate solution. In: International conference on calcium aluminate cements proceedings, pp 425–435
7. Wenda R, Kuzel HJ (1983) Proceedings of the 8th international congress on the chemistry of cement, Rio de Janeiro, vol 3, pp 31–38
8. Csetenyi J, Glasser FP (1993) Borate substituted ettringites. In: Material Research Society symposium proceedings, vol 294, pp 273–278
9. Bothe JV, Brown PW (1997) Phase formation in the system CaO – Al2O3–B2O3–H2O at 23 +/– 1 °C. J Hazard Mater B63:199–210
10. Casabonne JM (1987) Immobilisation des Radioelements Provenant de Dechets Nucléaires Sous forme de Composés Insolubles ou Par Réactions d'Echanges dans une Matrice de Beton. PhD Thesis, Bourgogne University, France
11. Roux C (1989) Conditionnement par des Liants Hydrauliques de Concentrats Boratés Radioactifs, PhD Thesis, Paris Sud-Orsay University, France

Chapter 19
The Effect of Supplementary Pulverised Fuel Ash on Calcium Aluminate Phosphate Cement for Intermediate-Level Waste Encapsulation

P. Swift, H. Kinoshita, and N.C. Collier

1 Introduction

In the UK, cement-based encapsulation has been established as the accepted practice for the immobilisation of low- and intermediate-level radioactive waste (LLW and ILW) [1, 2]. However, the high pH of the pore solution and availability of 'free-water',[1] in conventional ordinary Portland cement (OPC) systems causes corrosion of reactive metals such as aluminium [3, 4]. The corrosion reactions result in expansive reaction products and the generation of significant volumes of hydrogen gas, both of which are detrimental to the integrity of the solid wasteform. These metals make up a significant proportion of the UK waste inventory [5], so it is appropriate to develop alternative cementing systems which are compatible with these problematic reactive metals.

One such potential system is a type of acid–base phosphate-bonded cement, in which a cation-leachable powder acting as a base reacts with an acidic phosphate solution [6]. The neutralisation reactions result in rapid setting and the formation of reaction products, which act as cementitious binders. Sugama et al. [7, 8] synthesised cementitious systems for geothermal well applications by mixing calcium aluminate cement (CAC) with acidic phosphate-based solutions. These systems exhibited lower pH than the conventional OPC-based systems, and so have potential to reduce corrosion of problematic aluminium metal. Aluminium metal is passive in the pH range 4–8.5 according to the Pourbaix diagram [9].

[1] Water that is not chemically bound within the cement matrix, and is therefore available to react.

P. Swift (✉) • H. Kinoshita
Immobilisation Science Laboratory, Department of Materials Science and Engineering, The University of Sheffield, Sheffield, UK
e-mail: p.swift@shef.ac.uk

N.C. Collier
Immobilisation Science and Technology Team, National Nuclear Laboratory, Warrington, UK

F. Bart et al. (eds.), *Cement-Based Materials for Nuclear Waste Storage*,
DOI 10.1007/978-1-4614-3445-0_19, © Springer Science+Business Media New York 2013

Also, preliminary investigations by the authors have shown that corrosion of aluminium significantly decreases when pH < 10.5.

In a previous study by the authors [10], Secar 51—a type of CAC—was mixed with sodium polyphosphate-based solutions. The resulting cement pastes showed potential for waste encapsulation, exhibiting rapid setting (<8 min) and high strength development (>75 MPa after 24 h). However, the rapid setting and the significant heat output resulting from the setting reactions deemed the system unsuitable for industrial application. Specific processing criteria are defined by industry [Hayes M, Personal communication, National Nuclear Laboratory, 18/11/2010; Collier NC, Personal communication, National Nuclear Laboratory, 28/05/2010], which must be fulfilled before any system can be considered for LLW or ILW encapsulation. These criteria, referred from now on as 'plant acceptance tests', specify: the fluidity of the slurry, as measured by the Colflow test, should be >200 mm after 150 min (t_{150}) after mixing, the cement should exhibit bleed of <2 % by volume at 24 h after the mixing period; the initial and final setting time should be >4 h and <48 h, respectively. Industry also stipulates that cementitious matrices for waste encapsulation should develop compressive strength exceeding 7 MPa after 90 days curing.

The objective of the present study was to develop a cement formulation envelope which fulfils the processing and property criteria which are defined by industry. Pulverised fuel ash (PFA, otherwise known as fly ash) was considered to be an inert filler which would extend the setting time and reduce the heat output of calcium aluminate phosphate (CAP) cement [11]. Four systems were investigated, with varying proportions of PFA, and assessed in terms of (1) plant acceptance tests and pH of the wet cement slurries; (2) physio-mechanical properties, evaluated by compressive strength testing and porosity measurements; (3) phase compositions and microstructure of the various hardened cement formulations.

2 Experimental

A commercially available CAC (Secar 51, supplied by Kerneos UK) was used as the basic-solid reactant. The X-ray diffraction analysis (XRD) of this powder published previously [10] confirmed the major crystalline phase as monocalcium aluminate (CA), with gehlenite (C_2AS) and perovskite (CT) present in lesser amounts.[2] PFA (Cemex 450-S BS EN450-1, supplied by the National Nuclear Laboratory (NNL)) was used as the supplementary material, and XRD confirmed the presence of quartz and mullite as the primary crystalline phases in this powder. Details of the formulations investigated are given in Table 19.1. Based on the literature [12, 13], boric acid was used as a setting retarder. Sodium polyphosphate ($(NaPO_3)_n$, Acros Organics) and boric acid (H_3BO_3, 99.5+ %, Fisher Scientific)

Table 19.1 Details of the various formulations under investigation[a]

Formulation				Weight (%)					
f/s	*w/s*	*p/c*	*r/c*	CAC	PFA	$(NaPO_3)_n$	H_3BO_3	Water	Total
0.5	0.35	0.4	0.03	31.95	31.95	12.78	0.96	22.36	100
0.6	0.35	0.4	0.03	26.28	39.42	10.51	0.79	23.00	100
0.7	0.35	0.4	0.03	20.28	47.33	8.11	0.60	23.66	100
0.8	0.35	0.4	0.03	13.93	55.71	5.57	0.42	24.37	100

[a]Note: c = cement (CAC), f = filler (PFA), s = solid = $c + f$, p = phosphate (sodium polyphosphate), r = retarder (boric acid), w = water

powders were firstly dissolved in distilled water and left for at least 2 h on a roller-mixer to assure complete dissolution. The resultant solution was placed in the mixing bowl of a Kenwood bench-top planetary mixer set at speed 1 (ca. 150 rpm). Firstly PFA, then CAC powders were added to the solution over a ca. 5 min period, and mixed for a total of 10 min, followed by 10 min in a Silverson high shear mixer at 6,000 rpm.

Setting times were measured using an automatic Vicat measuring apparatus, based on BS EN 196-3:2005 + A1:2008 [14], with the end of the mixing period taken as $t = 0$ min (t_0). The fluidity of the cement slurries was determined at t_0 and after 150 min (t_{150}) using a standard slurry flow test, i.e. the Colflow test [Hayes M, Personal communication, National Nuclear Laboratory, 18/11/2010] in which a pre-set volume of cement slurry was released from a tundish into a channel, and the distance the slurry advanced down the channel was measured as the flow. The flow value was taken as the average of two Colflow tests. The pH of the wet cement slurries was measured at t_0, then every 30 min up to t_{150}. The compressive strengths were determined after a 7-day curing period as the mean of three 50 × 50 × 50 mm specimens, initially cured for 48 h at 20 °C and 95 % relative humidity (RH), and then demoulded and returned to the same curing conditions for the remainder of the curing period. The compressive strengths were measured using a Controls Automax5 automatic compressive strength testing machine. The specimens tested for compressive strength were further analysed by XRD, thermogravimetric analysis (TGA), scanning electron microscopy (SEM), and mercury intrusion porosimetry (MIP). To prepare these samples, the hydration reactions were arrested by immersing the crushed samples in acetone for 2 days and then drying in a vacuum desiccator for 7 days. XRD was performed using a Siemens D500 diffractometer using Cu K_α radiation over the range 5–65° 2θ, with a step size of 0.02° and scanning speed of 2° 2θ/min. TGA was performed with a Perkin Elmer Pyris 1 thermogravimetric analyser, with a heating profile of 10 °C/min over the temperature range 25–1,000 °C, under a nitrogen atmosphere. SEM was performed on polished surfaces of samples mounted in epoxy resin, using a Jeol JSM 6400 scanning electron microscope.

3 Results and Discussion

3.1 Plant Acceptance Tests and pH Measurements

Table 19.2 exhibits the results from the plant acceptance tests. Vicat results showed that the setting time increased with increasing PFA-to-solid ratio (*f/s*), and all formulations fulfilled the setting time requirements for both initial setting (>4 h) and final setting (<48 h). Fluidity, measured by the Colflow test, also fulfilled the minimum requirement of 200 mm at t_{150} and increased with increasing *f/s* ratio. It is well established that the incorporation of supplementary PFA increases the workability of OPC-based cementing systems [1, 15]. PFA consists of spherical particles, which increase the fluidity of cement slurries by the so-called ball bearing effect [16, 17]. Therefore, it is reasonable to assume that the same ball bearing effect is the principal mechanism responsible for increasing fluidity in CAP cementing systems with supplementary PFA. The four formulations studied exhibited no bleed after 24-h curing.

pH measurements of the wet cement slurries (Fig. 19.1) indicated that the pH of all systems lay between 9.1 and 10.0 at t_0, then gradually increased up to t_{150}, where they exhibited pH 10.2–10.6. At all times, pH increased with increasing *f/s* ratio. This reflected a decrease in the amount of phosphate and boric acid, which was available to participate in the acid–base setting reactions. The pH of studied systems was lower than that of an OPC-based system (pH 12.6–13.1), also shown in Fig. 19.1, indicating a potential as an alternative cementing system for the encapsulation of aluminium containing wastes.

3.2 Physio-mechanical Properties

The compressive strength tests demonstrated that the *f/s* ratio affected the mechanical properties of the hardened cement (Fig. 19.2a). Specifically, increasing the *f/s* ratio corresponded to an increase in porosity (Fig. 19.2b), and consequently a decrease in compressive strength. It is proposed that this increase in porosity was a result of the water-to-cement ratio (*w/c*) effectively increasing with *f/s* ratio. It is also suggested that the decrease in strength was a consequence of a lower CAC and phosphate

Table 19.2 Plant acceptance test results for PFA: CAC + $(NaPO_3)_n$ + H_3BO_3 mixed at $w/s = 0.35$, $p/c = 0.4$, r/c 0.03 and varying *f/s* ratio

	Setting time (h)		Colflow (mm)		
f/s	Initial	Final	t_0	t_{150}	Bleed (%)
0.5	4.5	<24	330	490	0
0.6	6.5	<24	510	720	0
0.7	7.0	<24	745	720	0
0.8	8.5	<24	850	870	0

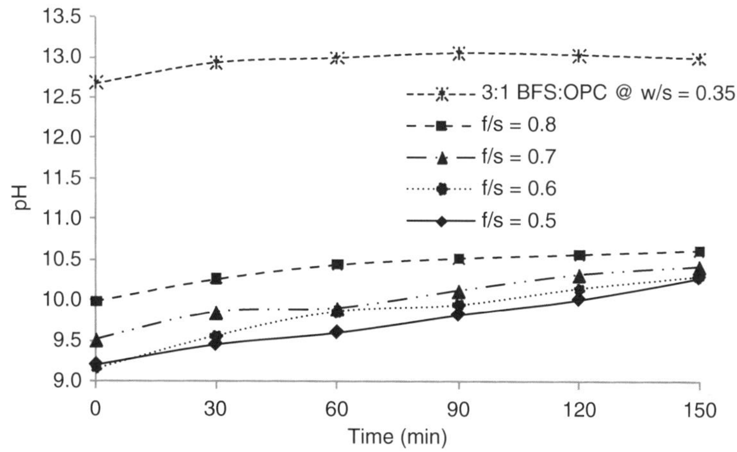

Fig. 19.1 pH measurements of wet cement slurries for PFA:CAC + $(NaPO_3)_n$ + H_3BO_3 mixed at $w/s = 0.35$, $p/c = 0.4$, $r/c = 0.03$ and varying f/s, and 3:1 BFS:OPC @ $w/s = 0.35$

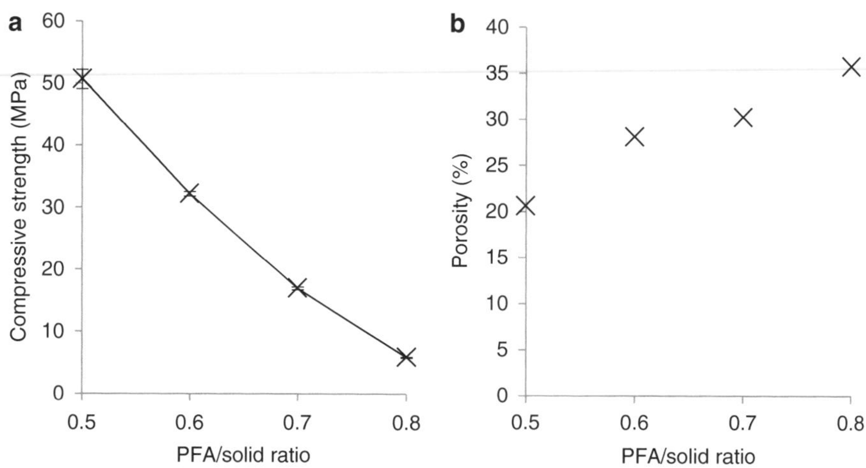

Fig. 19.2 Physico-mechanical results for PFA:CAC + $(NaPO_3)_n$ + H_3BO_3 mixed at $w/s = 0.35$, $p/c = 0.4$, $r/c = 0.03$ and varying f/s, after 7 days curing at 20 °C: (**a**) Compressive strength, (**b**) MIP

content, the reactions of which should form the binding matrix [10]. The $f/s = 0.8$ system attained a compressive strength of ca. 6 MPa after the 7-day curing period, which does not meet the 7 MPa minimum criterion required by industry. All other systems achieved compressive strengths greater than the minimum required after 7 days of curing.

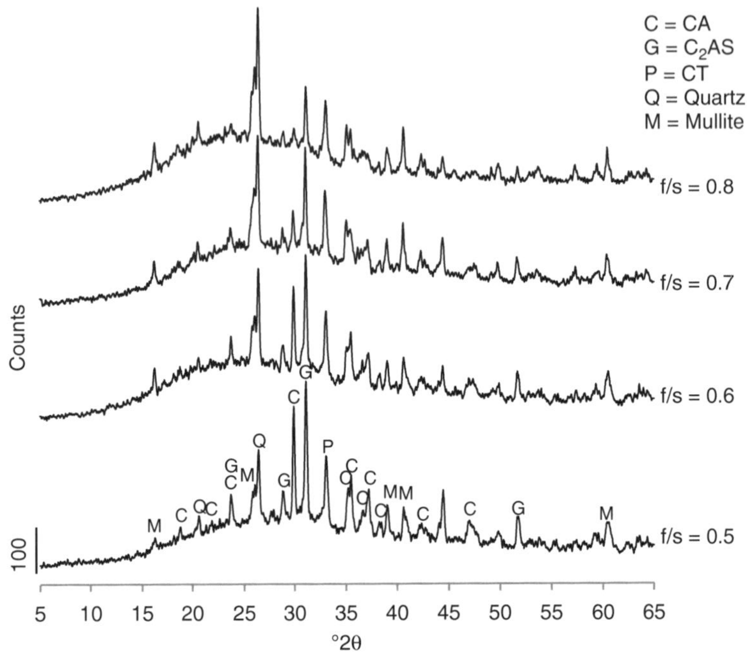

Fig. 19.3 XRD results for PFA:CAC + $(NaPO_3)_n$ + H_3BO_3 mixed at $w/s = 0.35$, $p/c = 0.4$, $r/c = 0.03$ and varying f/s, after 7 days curing at 20 °C

3.3 Phase Characterisation and Microstructure

The XRD results (Fig. 19.3) illustrate that the crystalline phase compositions of the various hardened cement pastes were very similar after 7 days curing. All systems exhibited the presence of CAC-related phases (CA, C_2AS, and CT), which was indicative of residual CAC phases that remained unreacted within the hardened cement paste. Reflections corresponding to quartz and mullite were also present from the PFA. As expected, the relative intensity of the CAC-related reflections appeared to decrease, and conversely those of the PFA appeared to increase, with increasing f/s ratio. The relative intensities of CA-related reflections decreased more significantly with increasing f/s ratio, compared to those of C_2AS and CT. This suggests that a higher proportion of the CA was consumed in these systems, which was thought to be attributed to the increasing w/c ratio.

The XRD results showed no indication of the crystalline CA-hydrate phases (CAH_{10}, C_2AH_8, and C_3AH_6), which form during the setting and strength development of CAC under normal hydration. As setting and strength development occurred, despite the apparent lack of CA-hydrate phases, it is suggested that an alternative reaction mechanism occurred, resulting in the formation of an amorphous binding

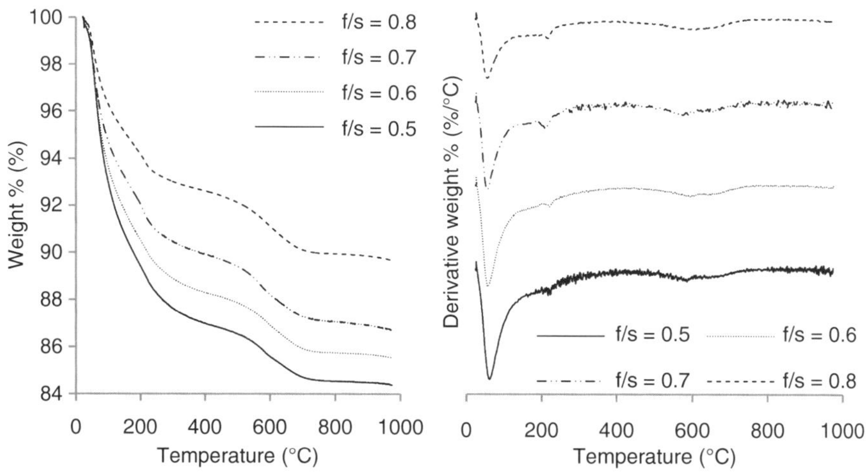

Fig. 19.4 TGA and DTG for PFA:CAC + $(NaPO_3)_n$ + H_3BO_3 mixed at $w/s = 0.35$, $p/c = 0.4$, $r/c = 0.03$ and varying *f/s*, after 7 days curing at 20 °C

phase. Similar results were obtained in the authors' previous study [10], which suggested that reactions between Secar 51 and polyphosphate-based solutions resulted in the formation of an amorphous phase, which acted as a binding matrix between the partially and un-reacted CAC particles, and was responsible for rapid setting and high strength development.

Results from TGA (Fig. 19.4) demonstrated that the total weight loss increased with decreasing *f/s*, suggesting an increase in the extent of reactivity with decreasing *f/s*. The principal weight loss below 120 °C, apparent from the corresponding differential thermogravimetric analysis (DTG) curves, may be due to the dehydration of the amorphous binding phases. Evaporation of residual acetone and free- and loosely bound water may also account for some of this weight loss. The minor weight losses, which occurred at ca. 200–220 °C and 550–700 °C, may be due to the dehydration of strätlingite, and the decomposition of calcite, respectively. Because neither strätlingite nor calcite were detected by XRD, if present they are only in small amounts. Strätlingite formation has been observed in CAC systems to which silica-rich supplementary materials were added [12, 18, 19]. The metastable CA-hydrate phases, which form during conventional hydration, react with silica to form strätlingite. It has also been proposed that after long curing periods, gehlenite may hydrate to form strätlingite [12, 20].

Figure 19.5 shows SEM micrographs using back scattered electron imaging (BSE) of two of the formulations investigated. Both micrographs showed particles of spherical PFA and unreacted CAC particles embedded within a matrix, which appeared to act as a binder between the particles. EDX analysis (not shown) of the apparent binding matrices exhibited a composition of calcium, aluminium, phosphorus, sodium, and oxygen. Sugama and Carciello [8] investigated sodium phosphate modified CAC systems, and suggested that a similar binding matrix consisted of sodium calcium orthophosphate ($NaCaPO_4 \times H_2O$) salt and hydrated

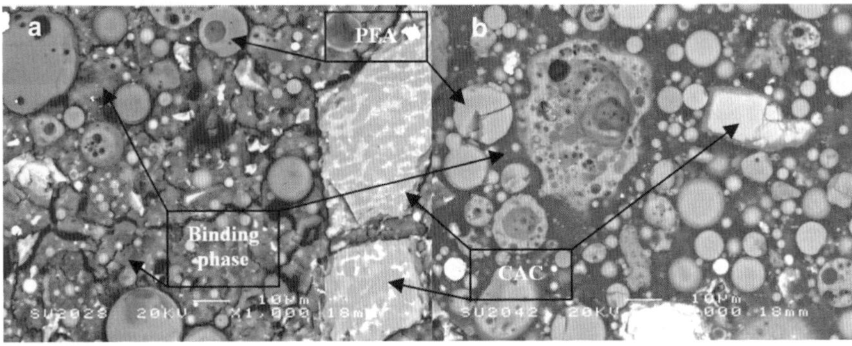

Fig. 19.5 SEM-BSE of polished surface samples of PFA:CAC + $(NaPO_3)_n$ + H_3BO_3 mixed at $w/s = 0.35$, $p/c = 0.4$, $r/c = 0.03$ and (**a**) $f/s = 0.5$ and (**b**) $f/s = 0.8$, after 7 days curing at 20 °C

alumina gel ($Al_2O_3.\times H_2O$). It is proposed that the binding matrix in the present study was a similar binary system consisting of $NaCaPO_4.\times H_2O$ and $Al_2O_3.\times H_2O$. It was apparent that the microstructure of the sample with $f/s = 0.5$ contained a higher proportion of unreacted CAC particles, and conversely that with $f/s = 0.8$ contained a higher proportion of PFA particles. This observation concurred with the results from XRD, which suggested that the proportion of unreacted CAC phases that remained in the hardened cement pastes decreased, and the proportion of PFA phases increased with increasing f/s ratio. Figure 19.5a also shows some cracks, the reason for which is not clear; they may have formed during samples preparation. They may also be thermally induced cracks which formed as a result of exothermic setting reactions. Systems at lower f/s ratio contained higher proportion of the reactive components, CAC and phosphate and such systems would have a higher exotherm due to the acid–base setting reactions.

4 Conclusions

All the systems investigated fulfilled the processing requirements defined by industry. Varying the f/s ratio had significant effects on various aspects of the systems. Both fluidity and setting time increased with f/s ratio. The porosity of the systems also increased with increasing f/s ratio due to thc increasing effective w/c ratio. Consequently, compressive strength decreased with increasing f/s ratio. Apart from the system with $f/s = 0.8$, all systems developed compressive strength in excess of the 7 MPa requirement, after 7 days curing.

The crystalline phase compositions of the studied systems were very similar, with reflections corresponding to CAC- and PFA-related phases apparent. It appears that a higher proportion of CA was consumed as the f/s ratio increased, which was likely due to the increasing w/c ratio. The crystalline CA-hydrate phases were not formed, nor were any other new crystalline phases, suggesting that the reactions products were amorphous. Using SEM-BSE, the amorphous phases were observed to be acting as a binding matrix between the partially and un-reacted CAC, and the spherical PFA particles, and were responsible for setting and strength development. It was proposed that this matrix was a binary system consisting of $NaCaPO_4.\times H_2O$ and $Al_2O_3.\times H_2O$.

The present study identified a formulation envelope for calcium aluminate phosphate cements, which fulfils the processing and property criteria defined by industry for LLW and ILW encapsulation. Along with our previous study [10], the work presented here indicated that calcium aluminate phosphate cement-based systems have potential for encapsulating problematic LLW and ILW in the UK.

Acknowledgments Thanks are given to the Immobilisation Science Laboratory (ISL) of the University of Sheffield. Special thanks are given to Dr. Claire Utton of the ISL, Dr. Martin Hayes of the National Nuclear Laboratory (NNL), and Dr. Neil Milestone of Milestone and Associates Ltd., New Zealand for their helpful discussions and support. The authors gratefully acknowledge the Engineering and Physical Sciences Research Council and NNL for funding.

References

1. Glasser FP (1992) Progress in the immobilization of radioactive wastes in cement. Cement Concr Res 22:201–216
2. Wilding CR (1992) The performance of cement based systems. Cement Concr Res 22 (2–3):299–310
3. Setiadi A et al (2006) Corrosion of aluminium and magnesium in BFS composite cements. Adv Appl Ceram 105:191–196
4. Milestone NB et al (2006) Reactions in cemented nuclear waste forms – the need for a toolbox of different cement types. In: Materials Research Society symposium proceedings, vol 932
5. The 2007 UK Radioactive Waste Inventory (2007) Department for Environment, Food and Rural Affairs and Nuclear Decommissioning Authority, pp 1–140
6. Sugama T, Allan M, Hill JM (1992) Calcium phosphate cements prepared by acid-base reaction. J Am Ceram Soc 75(8):2076–2087
7. Sugama T, Carciello NR (1991) Strength development in phosphate-bonded calcium aluminate cements. J Am Ceram Soc 74(5):1023–1030
8. Sugama T, Carciello NR (1995) Sodium phosphate-derived calcium phosphate cements. Cement Concr Res 25(1):91–101
9. Setiadi A (2006) Corrosion of metals in composite cement. In: Department of Engineering Materials, The University of Sheffield, Sheffield
10. Swift P, Kinoshita H, Collier NC, Utton CA (2012) Phosphate modified calcium aluminate cement for radioactive waste encapsulation. Advances in Applied Ceramics
11. Sharp JH et al (2003) Cementitious systems for encapsulation of intermediate level waste. In: Proceedings of ICEM '03: the 9th international conference on radioactive waste management and environmental remediation, Oxford, England
12. Scrivener KL, Campas A (2006) Calcium aluminate cements, Chapter 13. In: Hewlett PC (ed) Lea's chemistry of cement and concrete. Butterworth-Heinemann, Oxford, pp 711–782
13. Sugama T, Carciello NR, Gray G (1992) Alkali carbonation of calcium aluminate cements: influence of set-retarding admixtures under hydrothermal conditions. J Mater Sci 27 (18):4909–4916
14. Methods of testing cement (2005) Determination of setting times and soundness, BS EN 196-3:2005+A1:2008, The British Standards Institution

15. Atkins M, Glasser FP (1992) Application of Portland cement-based materials to radioactive waste immobilization. Waste Manage 12(2–3):105–131
16. Megat Johari MA et al (2011)Influence of supplementary cementitious materials on engineering properties of high strength concrete. Construct Build Mater 25(5):2639–2648
17. Lee SH et al (2003) Effect of particle size distribution of fly ash-cement system on the fluidity of cement pastes. Cement Concr Res 33(5):763–768
18. Chatterjee AK (2002) Special cements, Chapter 6. In: Bensted J, Barnes P (eds) Structures and performance of cements. Spon Press, London
19. Majumdar AJ, Edmonds RN, Singh B (1990) Hydration of Secar 71 aluminous cement in presence of granulated blast furnace slag. Cement Concr Res 20(1):7–14
20. Bensted J (2002) Calcium aluminate cements, Chapter 4. In: Bensted J, Barnes P (eds) Structure and performance of cements. Spon Press, London

Chapter 20
Rheological Behavior of Alkali-Activated Metakaolin During Geopolymerization

A. Poulesquen, F. Frizon, and D. Lambertin

1 Introduction

Geopolymers are a class of largely X-ray amorphous three-dimensional aluminosilicate binder materials, synthesized by reaction of an aluminosilicate powder with a concentrated alkali metal silicate or hydroxide solution [1]. Geopolymerization is a complex process involving dissolution of solid reactants, hydrolysis of the dissolved species, and condensation of the geopolymer itself [2] and can be described in two steps. First, a dissolution step, during which reactive aluminate and silicate species are brought into solution. The chemical driver of this step is the action of hydroxide ion on aluminosilicate that lead to monomeric, and probably partially deprotonated, silicates and aluminates [3]. The development of microstructure following the hydrolysis/dissolution steps arises out of condensation mechanisms between isolated monomerics units. The resulting formation of two- and three-dimensional linkages via oxygen bridges eliminates water, while alkali is not regenerated. The types of network formed are very dependent on the chemistry of the system [3, 4].

Although dynamic rheological experiments may provide viewpoint on geopolymerization reactions, see Bénard et al. [5] for example on cement-based materials, no such study has been carried out on geopolymers. The rheological behavior of geopolymers was frequently determined using a viscometer, i.e., in flowing mode, because it is of primary importance to commercial applications in the

A. Poulesquen (✉) • D. Lambertin
Commissariat à l'Energie Atomique et aux Energies Alternatives, DEN/MAR/DTCD/SPDE, Bagnols-sur-Cèze, France
e-mail: arnaud.poulesquen@cea.fr

F. Frizon
Commissariat à l'Energie Atomique et aux Energies Alternatives, DEN/MAR/DTCD/SPDE, Bagnols-sur-Cèze, France

Laboratoire d'Etude de l'Enrobage des Déchets, French Atomic Energy and Alternative Energies Commission, Bagnols sur Cèze Cedex, France

F. Bart et al. (eds.), *Cement-Based Materials for Nuclear Waste Storage*,
DOI 10.1007/978-1-4614-3445-0_20,

field of civil engineering [6], or wastes immobilizing materials [7–9]. Published work has in particular addressed the determination of the yield stress and Newtonian viscosity [10] by simulating the flow curves with a Bingham or Herschel-Bulkley model [11]. Criado et al. [12] showed that adding a superplasticizer diminishes the viscosity and the yield stress of a fly ash-based geopolymer.

In this study we propose to consider geopolymerization reactions by observation of the viscoelastic properties of the geopolymer using dynamic rheology in the linear viscoelastic domain. The linearity range was first determined and the main viscoelastic parameters were evaluated. The effect of the activation solution composition, which is of primary importance on the geopolymerization reaction [3] was investigated. Finally, as geopolymers are usually synthesized at or slightly above room temperature, the temperature dependence of the reactions with different materials was also assessed and the activation energies were determined.

2 Experimental

2.1 Material

In order to focus on a model system and avoid the precipitation of calcium silicate hydrate phases mixed with geopolymeric materials [13], the choice was made to synthesize metakaolin-based geopolymer paste. This product was purchased under the brand name of Pieri Premix MK from Grace Construction Products. The chemical composition of metakaolin determined by X-ray Fluorescence (XRF) can be found in Table 20.1. X-ray Diffraction (XRD) analysis showed that the metakaolin contained anatase, kaolinite, and quartz as impurities.

Waterglass activating solutions were prepared by simultaneously dissolving alkali hydroxide (NaOH, KOH: Prolabo, Rectapur, 98 %) pellets and amorphous silica provided by Rhodia (Tixosil 331) and Degussa (Aerosil 380) in Milli-Q water. Solutions were then stirred for 15 h. Two types of silica were used in the synthesis of the alkali silicate solution: fumed silica produced at high temperature (Aerosil) and precipitated silica (Tixosil). The different synthesis processes form different particle size fractions and different silanol and siloxane site densities, and thus possibly different reactivity properties. The physical and chemical characteristics of the solids used are listed in Table 20.2.

All the geopolymers were prepared with the same overall formulation, with an $Al:Si:M:H_2O$ molar ratio of 1:1.8:1:12. The only differences among the samples were therefore the alkali metal (M = Na or K) and the silica (Aerosil or Tixosil) used (Table 20.3).

Table 20.1 Chemical composition of metakaolin used in this study

	SiO_2	Al_2O_3	CaO	Fe_2O_3	TiO_2	K_2O	Na_2O	MgO	L.o.I[a]
Weight (%)	54.4	38.4	0.10	1.27	1.6	0.62	<0.2	<0.2	1.90

[a]Loss on ignition

Table 20.2 Characteristics of solids used to synthesize the geopolymers

Solid	Specific surface area measured by the BET method (m^2/g)	Median diameter determined by laser particle size analysis (d_{50}) (μm)	Number of surface Si-OH sites measured by the Sears method [25] (OH/nm^2)
Metakaolin	19.9 ± 0.8	9.7 ± 0.7	
Tixosil 331	172 ± 2.5	3.5 ± 0.3	3.3
Aérosil 380	378 ± 4.3	0.7 ± 0.2	1.6

Table 20.3 Characteristics and formulation of geopolymers studied

Sample	Aluminosilicate source	Alkali activator(MOH)	Silica
1	Metakaolin	NaOH	Aerosil 380
2	Metakaolin	KOH	Aerosil 380
3	Metakaolin	NaOH	Tixosil 38
4	Metakaolin	KOH	Tixosil 38

2.2 Rheological Measurements

Dynamic rheological measurements were performed using a controlled stress rheometer (AR1000, TA Instruments) operated in controlled strain mode. Measurements were performed in unconventional geometry to facilitate mixing, using a bowl and a helical 316 L steel ribbon impeller calibrated according to the procedure described in the literature [14]. This set of geometry was designed in order to avoid the evaporation (and consequently the geopolymer paste to dry), the phase separation during the curing time, and so to limit the wall effects.

The geopolymer paste was prepared manually outside the rheometer and then introduced into the rheometer bowl after mixing. Transferring the geopolymer to the measurement cell generates residual stresses that must be relaxed prior to each rheological measurement. This is done before each test by applying a strain of 0.05 and pulsation at a rate of 5 rad s^{-1} [5].

Since the purpose of this study is to determine the variation over time of the viscoelastic parameters according to the geopolymer composition, it is important to ensure the repeatability of the tests and to determine the linear viscoelasticity range in order to prevent structural decomposition of the geopolymerized network.

Figure 20.1 shows the strain sweep for 2 frequencies (1 and 10 rad s^{-1}) including a repeatability test for the 10 rad s^{-1} pulsation, which was carried out by integrating 5 or 10 points per order of magnitude. Given the relatively fast geopolymerization kinetics, the acquisition of 10 measurement points per order of magnitude doubles the measurement time, which can be problematic for some systems. Nevertheless, the repeatability was very satisfactory.

The linear viscoelasticity region is the region in which measurements are nondestructive to the geopolymerized network when a critical strain γ_C is applied. γ_C was generally taken as the strain value at the storage modulus equal to 90 % of the plateau modulus (0.9 G'). For the geopolymers synthesized here, the limit of the

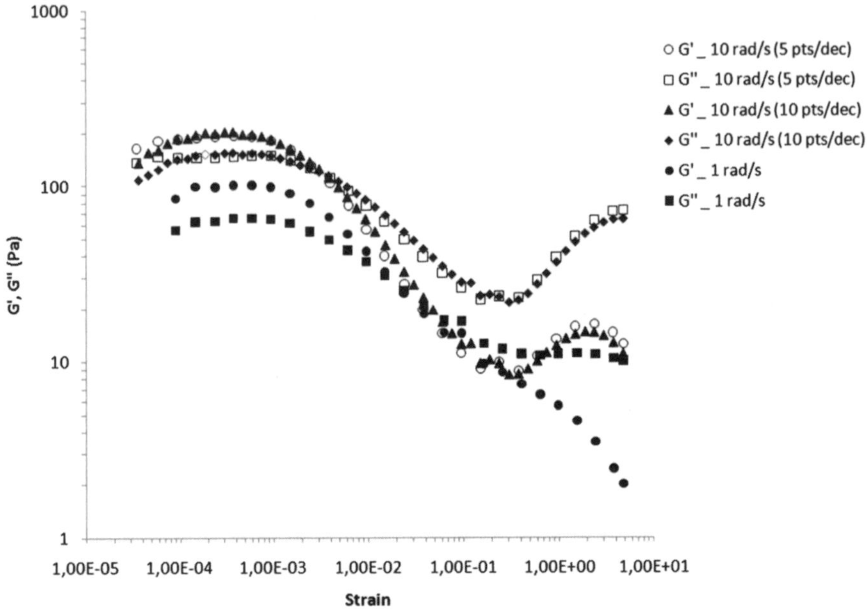

Fig. 20.1 Reproducibility of strain sweep test at $\omega = 10$ rad s^{-1} for Metakaolin activated by NaOH and strain sweep test at $\omega = 1$ rad s^{-1}: ($T = 25$ °C)

linear viscoelasticity range is about 1.5×10^{-3}. This value is slightly higher than what can be obtained in conventional Portland cement where the strain limit is about 0.03 % [5, 15].

In Fig. 20.1, the yield stress (σ_0) value can also be obtained when $G' = G''$. Indeed, when G' is higher than G'' the material is considered as mostly a solid and on the other side when G' becomes lower than G'', the material flows.

$$\sigma_0 = G* \cdot \gamma_{G'=G''} = (G'^2 + G''^2)^{0.5} \cdot \gamma_{G'=G''}$$

The yield stress obtained is ranged between 0.5 and 1 Pa whatever the material studied in this chapter.

After the preconditioning step, all the experiments were therefore performed with a strain of 2×10^{-4} and at a frequency of 1 rad s^{-1}. The elastic modulus (G') and viscous modulus (G'') and the loss tangent defined by $\tan\delta = G''/G'$ were analyzed during geopolymerization.

3 Results and Discussion

3.1 Overall Evolution of Viscoelastic Parameters

Figure 20.2 shows the evolution of the rheological parameters (G', G'', and $\tan\delta$) during a geopolymerization test and until the reaction is complete (G' is constant

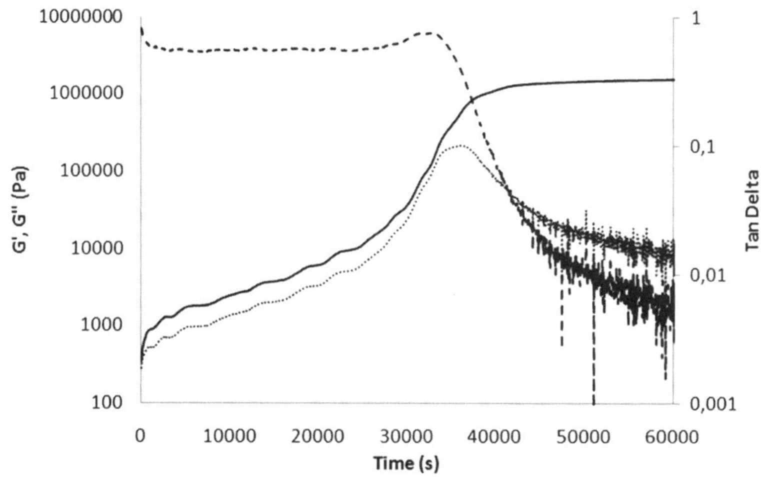

Fig. 20.2 Viscoelastic parameters versus time at ambient temperature ($T = 19.6$ °C) for geopolymer made from sodium hydroxide and Tixosil 38 silica

and G'' and tanδ tend to zero). The elastic modulus (G') quickly exceeds the viscous modulus, indicating the existence of interactions between the constituents and the predominance of the solid properties of the materials even before setting. Both moduli exhibit several levels during geopolymerization followed by an increase of G' and G''. The geopolymerization reactions involve mainly a dissolution/polycondensation mechanism [16, 17]. During the first stage, the dissolution of the solid aluminosilicate source by the alkaline solution results in the appearance in solution of an aluminum–silicon monomer [18]. In the second stage, these dissolved species react to form oligomers of different sizes, which undergo polycondensation reactions to form a three-dimensional aluminosilicate network [19–21]. The numerous steps observed on the G' and G'' curves during geopolymerization could therefore be related to the first stage, i.e., to a metakaolin dissolution mechanism, whereas the successive increases could be attributed to the formation of oligomers. The two mechanisms of dissolution and oligomer formation thus occur repeatedly over time.

The loss tangent is the ratio between the energy dissipated in the form of heat and the stored energy. Since G' varies more rapidly than G'', the value of the tangent diminishes and is always less than 1. At the end of the experiment, a tanδ peak appears as G'' increases faster than G'. The effects of energy dissipation in the form of heat then outweigh the stored energy associated with interactions between constituents. Various dissipation processes, including the formation of large oligomers or the beginning of polycondensation into a three-dimensional network, may account for this phenomenon, although it is difficult at this time to clearly identify the mechanism(s) responsible. Finally, the experiment was terminated when G' reached a constant value and when G'' and tanδ tended to zero, which means from a rheological point of view that a solid is formed. In the following figures of this chapter, the test was stopped just after the tanδ peak is reached to prevent deformation of the test geometry. Anyway, the maximum of the loss tangent is thus a precursor criterion of the vitreous state transition. Moreover, as discussed below, the evolution of tanδ depends on the nature of the alkali activator (NaOH or KOH).

3.2 *Effect of Silica*

It has been reported in the literature that sodium hydroxide activation solutions ensure better dissolution of the aluminosilicate source than solutions of potassium hydroxide [22–24], which is nevertheless a stronger base. This paradox can be explained by assuming silica is at chemical equilibrium in solution. Solution enrichment in silicate species arises from two separate reactions.

First, alkali hydroxides diffuse to the surface of the silicate particles and interact with them to ionize the surface silanol groups according to the following reaction:

$$\equiv \mathrm{Si} - \mathrm{OH} + \mathrm{OH}^-\mathrm{M}^+ \rightarrow \equiv \mathrm{Si} - \mathrm{O}^-\mathrm{M}^+ + \mathrm{H_2O} \quad (20.1)$$

During this process, the ionization rate depends on the basicity of the alkali ion, i.e., $K^+ > Na^+$.

The second reaction consists in breaking siloxane bonds in the silicate source, resulting in its dissolution:

$$\equiv \mathrm{Si} - \mathrm{O} - \mathrm{Si} \equiv +2(\mathrm{OH}^-\mathrm{M}^+) \rightarrow 2(\equiv \mathrm{Si} - \mathrm{O}^-\mathrm{M}^+) + \mathrm{H_2O} \quad (20.2)$$

During this process, the dissolution rate depends on the degree of hydration of the cation, i.e., $K^+ < Na^+$. This reaction appears to become predominant at high SiO_2/M_2O ratios.

Sodium hydroxide thus ensures better dissolution of the aluminosilicate source than potassium hydroxide, although the latter is a stronger base. To evaluate the impact of this reactivity difference on rheological behavior, geopolymers were synthesized from two types of silica (Aerosil 380 and Tixosil 38) to assess the effect of the number of silanol and siloxane sites on the initial reactivity of the materials.

Taken as a whole, the curves presented on Fig. 20.3a, b are similar to those described on Fig. 20.2. Indeed, the different levels, on both moduli, during geopolymerization are present (even if it seems noisier because the curves are plotted with dots) and at the end of the test a tanδ peak is observed.

For geopolymers made from potassium hydroxide (Fig. 20.3a), it appears that the nature of the silica does not strongly affect the rheological behavior since the curves are practically superimposed. However, a minor difference was measurable at the end of the experiment when the rate appeared to accelerate for the activation solution containing Tixosil 38 silica (Fig. 20.3a). In fact, the geopolymers formed from Tixosil 38 stiffened faster than those produced from Aerosil. As potassium

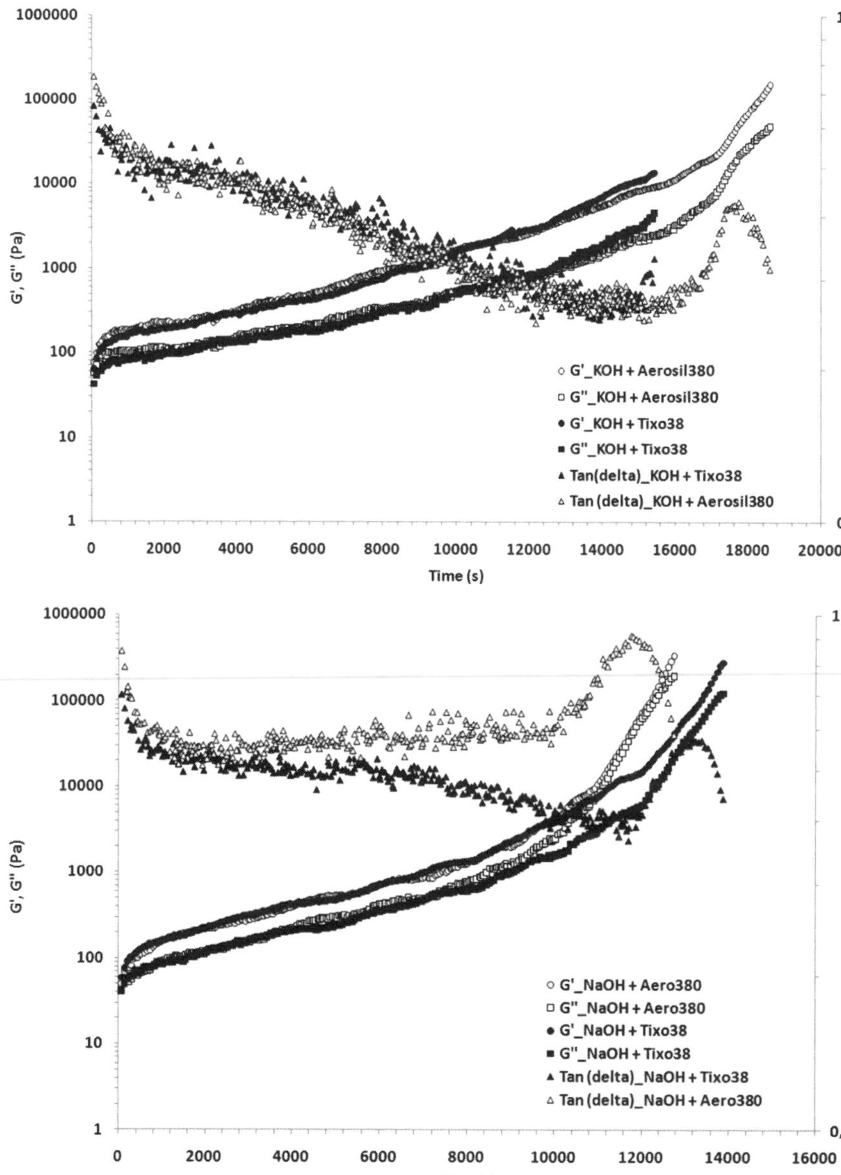

Fig. 20.3 Effect of silica grade on viscoelastic moduli for (**a**) geopolymer made from KOH; (**b**) geopolymer made from NaOH: (T = 25 °C)

favors the ionization of silanol sites more than the breaking of siloxane bonds, silica compounds with the greatest silanol surface density are the most reactive.

Conversely, when geopolymerization is activated by sodium hydroxide, the impact of the nature of the silica is more obvious, especially with regard to the evolution of

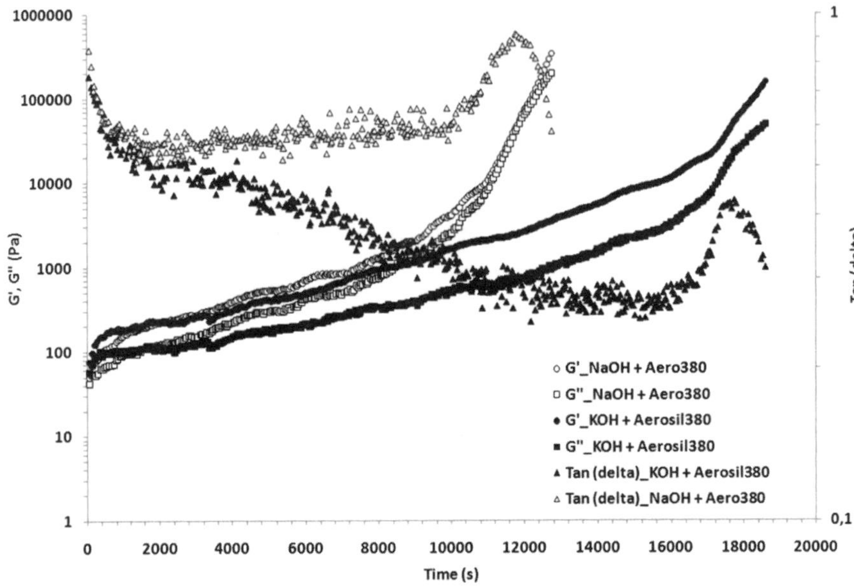

Fig. 20.4 Evolution of the elastic modulus (G') and viscous modulus (G'') and $\tan\delta$ according to alkali activation solution composition: ($T = 25\ ^\circ C$)

$\tan\delta$ (Fig. 20.3b). The silica with the largest number of siloxane sites and the smallest particle size distribution (Aerosil 380) creates a constant difference between G' and G'' ($\tan\delta$ is constant) but also reaches the maximum $\tan\delta$ more quickly. In the case of a sodium hydroxide activation solution, geopolymers formed from Aerosil stiffen more rapidly than the others: since sodium promotes silica dissolution by breaking siloxane bonds rather than by ionizing silanols, the silica containing the most siloxane groups proved to be the most reactive.

3.3 Effect of Alkali Activator

The alkali activator is known to play a major role in the kinetics of geopolymerization. Sodium hydroxide thus ensures better dissolution of the aluminosilicate source (MK) than potassium hydroxide, although the latter is a stronger base. Lee and Van Deventer [24] confirmed that the dissolution rate of the aluminosilicate source depends on the activation solution composition.

Figure 20.4 confirms that sodium hydroxide accelerates the geopolymer curing kinetics compared with potassium hydroxide, and that the flat portions of the viscoelastic moduli are shorter.

Due to their different charge densities, the small ions (Na^+) strongly attract and retain their hydration layer of water molecules, while larger alkali metal cations (K^+) do not have such a strong association with water: they may therefore associate more easily with the negative charge of the silicate, neutralizing the charge and stabilizing the suspension. Consequently, sodium ions are less likely than potassium ions to disrupt the condensation reaction. Moreover, tanδ versus time shows that the elastic component (G') varies more rapidly than the viscous component (G'') in the case of potassium hydroxide. This suggests that the geopolymer structure is more rigid when activated by KOH, forming larger oligomers or favoring the connectivity of the tetrahedral network. In other words, if the alkaline activation solution is potassium hydroxide, the geopolymerization kinetics are slower but with much stronger interactions between constituents ($\tan\delta_{KOH} < \tan\delta_{NaOH}$). These results are consistent with those of Xu et al. [16], which show that the compressive strength is 42 % higher when the geopolymer is synthesized from KOH.

3.4 Effect of Temperature

The effect of temperature on the geopolymerization kinetics was investigated by modifying the alkaline activation solution. Figures 20.5 and 6 show the evolution of the viscoelastic properties versus time at four different temperatures for both types of geopolymer. Regardless of the temperature and the material studied, the same trends are observed:

- Elastic behavior always predominates.
- Plateaus are always observed, but appear to be less numerous at higher temperatures. They are probably less likely to be detected because of the high rate of geopolymerization compared with the acquisition time.
- A peak is also observed for tanδ regardless of the measurement temperature (Figs. 20.5b and 20.6b).

The temperature dependence of the geopolymerization reactions may then be estimated by plotting an Arrhenius diagram showing the time necessary to reach the maximum tangent δ for a sodium- or potassium-based system (indicated by arrows in Figs. 20.5b and 20.6b) versus the temperature (Figs. 20.7 and 20.8, respectively). The activation energies are indicated in Table 20.4. The activation energy values are practically identical although the kinetics are very different for geopolymers made from sodium hydroxide and potassium hydroxide. This suggests that the geopolymerization mechanisms are similar for both types of alkali and that the process consists of successive stages of dissolution followed by the formation of oligomers whose size depends on the nature of the alkali activator, ultimately leading to the creation of a three-dimensional aluminosilicate network.

It is also interesting to note that the loss tangent decreases more rapidly as the temperature rises (this phenomenon is more pronounced with KOH Fig. 20.6b). The formation of a three-dimensional "gel" architecture is thus favored and accelerated at high temperature.

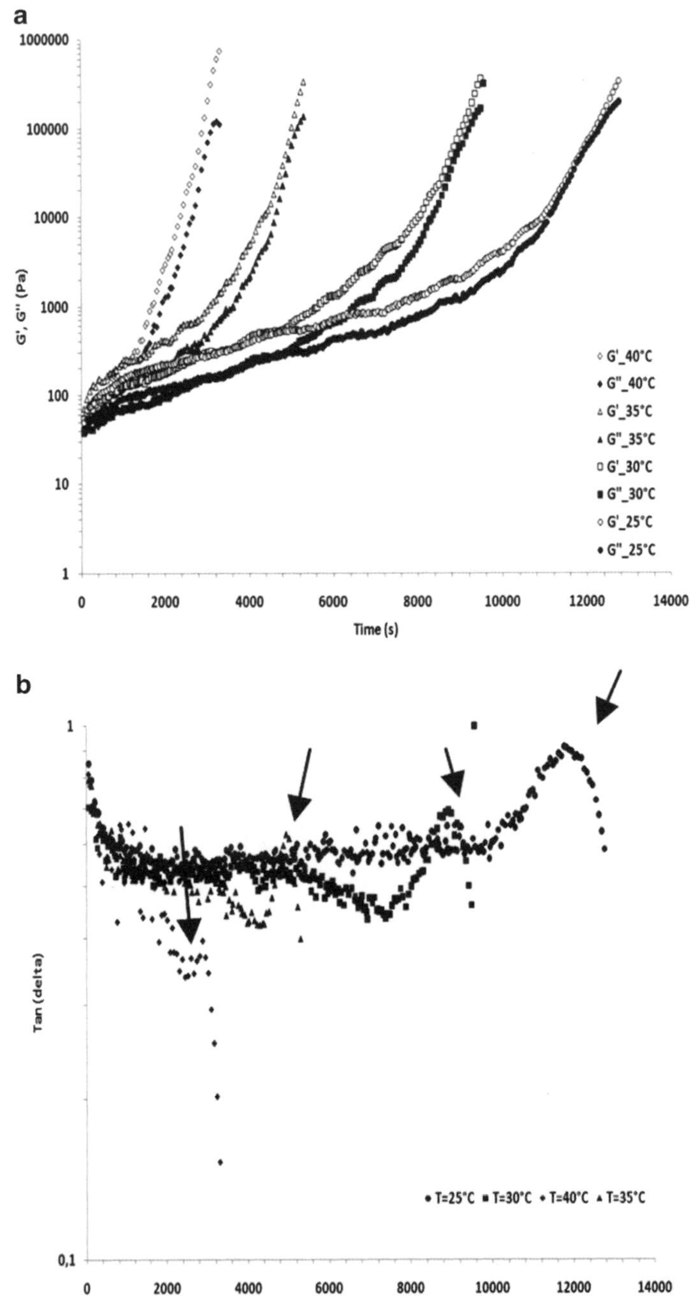

Fig. 20.5 Effect of temperature on (**a**) elastic modulus (G') and viscous modulus (G'') and (**b**) on $\tan\delta$ for NaOH + Aerosil 380 system

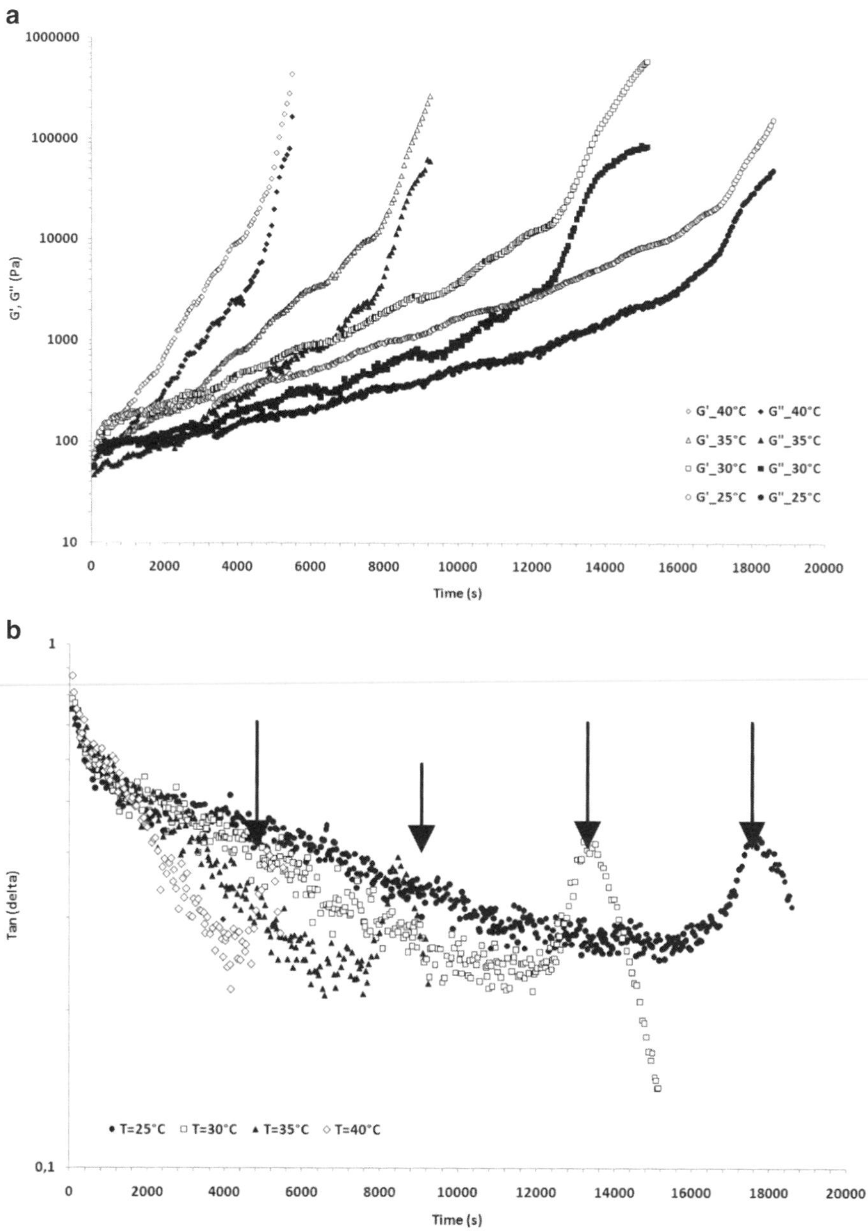

Fig. 20.6 Effect of temperature on (**a**) elastic modulus (G') and viscous modulus (G'') and (**b**) on $\tan\delta$ for KOH + Aerosil 380 system

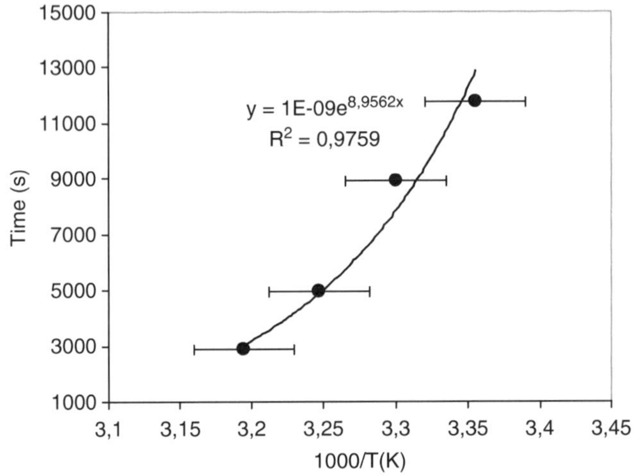

Fig. 20.7 Time to reach maximum tanδ versus temperature for NaOH + Aerosil 380 system

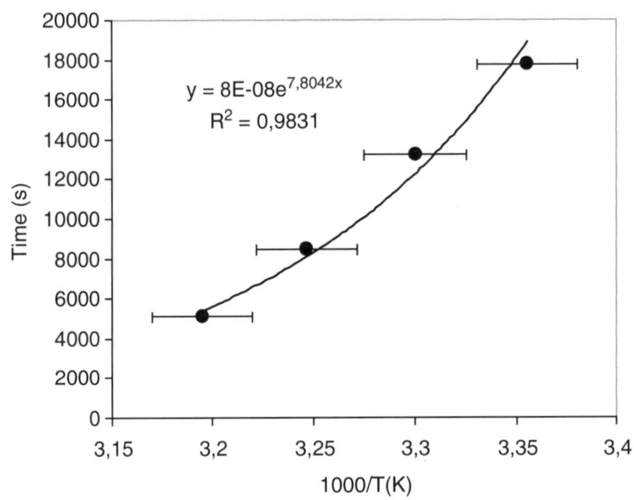

Fig. 20.8 Time to reach maximum tanδ versus temperature for KOH + Aerosil 380 system

Table 20.4 Activation energies of geopolymerization reactions

Geopolymer	Activation energy (kJ/mol)
NaOH + Aerosil 380	74.5
KOH + Aerosil 380	64.8

4 Conclusion

The evolution of the viscoelastic parameters of various geopolymers was evaluated by dynamic rheology. In general, the elastic modulus quickly exceeds the viscous modulus regardless of the geopolymer studied. During geopolymerization, several steps appear on the curves of the G' and G'' moduli, followed by a subsequent rise in both curves. These two characteristics are related respectively to an aluminosilicate matrix dissolution mechanism and to the formation of oligomers whose size depends on the alkaline activation solution used (NaOH or KOH). The geopolymerization kinetics is faster with sodium hydroxide, but with weaker interactions between constituents and/or more limited formation of oligomers. The loss tangent diminishes more rapidly in the case of KOH, which shows that it favors the formation of a three-dimensional architecture, probably due to greater oligomer connectivity: this type of geopolymer is therefore more rigid. The influence of the type of silica used is less obvious than that of the alkali activator, but it appears that when sodium hydroxide is used to activate the geopolymer, the effects are more pronounced, i.e., the smaller the particle size distribution, the faster the curing kinetics.

The maximum loss tangent, tanδ, was also used as a criterion to determine the temperature dependence of the geopolymers synthesized. This criterion is also a precursor of the transition to the vitreous state. The activation energies determined for the geopolymers synthesized with potassium hydroxide or sodium hydroxide are relatively close, showing that the mechanisms involved are similar regardless of the alkali activator.

Acknowledgments The authors are grateful to J. Beri, who performed most of the rheological measurements during her internship, and to A. Blachère and A. Gérenton for preparing the geopolymers.

References

1. Davidovits J (1991) Geopolymers – inorganic polymeric new materials. J Therm Anal 37:1633–1656
2. Fernandez-Jimenez A, Palomo A, Criado M (2005) Microstructure development of alkali-activated fly ash cement: a descriptive model. Cement Concr Res 35(6):1204–1209
3. Buchwald A, Zellmann H-D, Kaps C (2011) Condensation of aluminosilicate gels-model system for geopolymer binders. J Non Cryst Solids 357:1376–1382
4. Prud'homme E, Michaud P, Joussein E, Clacens JM, Rossignol S (2011) Role of alkaline cations and water content on geomaterial foams: monitoring during formation. J Non Cryst Solids 357:1270–1278
5. Desbats Le Chequer C, Frizon F (2011) Impact of sulfate and nitrate incorporation on potassium- and sodium-based geopolymers: geopolymerization and materials properties. J Mater Sci 46:5657–5664
6. Bénard P, Garrault S, Nonat A, Cau-Dit-Coumes C (2005) Hydration process and rheological properties of cement pastes modified by orthophosphate addition. J Eur Ceram Soc 25:1877–1883

7. Van Deventer JSJ, Provis JL, Duxson P, Brice DG (2010) Chemical research and climate change as drivers in the commercial adoption of alkali activated materials. Waste Biomass Valor 1:145–155
8. Palomo A, Palacios M (2003) Alkali-activated cementitious materials: alternative matrices for the immobilisation of hazardous wastes – part II. Stabilisation of chromium and lead. Cement Concr Res 33(2):289–295
9. Xu JS, Zhou YL, Chang Q, Qu HQ (2006) Study on the factors of affecting the immobilization of heavy metals in fly ash-based geopolymers. Mater Lett 60(6):820–822
10. Berger S, Frizon F, Joussot-Dubien C (2009) Formulation of caesium based and caesium containing geopolymers. Adv Appl Ceram 108:412–417
11. Palacios M, Houst YF, Bowen P, Puertas F (2009) Adsorption of superplasticizer admixtures on alkali-activated slag pastes. Cement Concr Res 39:670–677
12. Palomo A, Banfill PFG, Fernandez-Jimenez A, Swift DS (2005) Properties of alkali-activated fly ashes determined from rheological measurements. Adv Cement Res 17:143–151
13. Criado M, Palomo A, Fernandez-Jimenez A, Banfill PFG (2009) Alkali activated fly ash: effect of admixtures on paste rheology. Rheol Acta 48:447–455
14. Yip CK, Lukey GC, Van Deventer JSJ (2005) The coexistence of geopolymeric gel and calcium silicate hydrate at the early stage of alkaline activation. Cement Concr Res 35 (9):1688–1697
15. Ait-Kadi A, Marchal P, Choplin L, Chrissemant AS (2002) Quantitative analysis of mixer-type rheometers using the Couette analogy. Can J Chem Eng 80:1166–1174
16. Nachbaur L, Mutin JC, Nonat A, Choplin L (2001) Dynamic mode rheology of cement and tricalcium silicate pastes from mixing to setting. Cement Concr Res 31:183–192
17. Xu H, Van Deventer JSJ (2003) The effect of alkali metals on the formation of geopolymeric gels from alkali-feldspars. Colloids Surf A Physicochem Eng Asp 216(2003):17–44
18. Phair JW, Van Deventer JSJ (2002) Effect of the silicate activator pH on the microstructural characteristics of waste-based geopolymers. Int J Miner Process 66:121–143
19. Lee WKM, Van Deventer JSJ (2002) Structural reorganization of class F fly ash in alkaline silicate solutions. Colloids Surf A Physicochem Eng Asp 211:49–66
20. Xu H, Van Deventer JSJ (2000) The geopolymerisation of alumino-silicate minerals. Int J Miner Process 59:247–266
21. Palomo A, Alonso S, Fernandez Jiminez A, Sobrados I, Sanz J (2004) Alkaline activation of fly ashes: NMR study of the reaction products. J Am Ceram Soc 87(6):1141–1145
22. Babushkin VI, Matveyev GM, Mchedlov-Petrossyan OP (1985) Thermodynamics of silicates. Spinger, Berlin, pp 276–281
23. Hajimohammadi A, Provis JL, Van Deventer JSJ (2010) Effect of alumina release rate on the mechanism of geopolymer gel formation. Chem Mater 22:5199–5208
24. Li C, Sun H, Li L (2010) A review: the comparison between alkali-activated slag (Si plus Ca) and metakaolin (Si plus Al) cements. Cement Concr Res 40:1341–1349
25. Xu H, Van Deventer JSJ (2002) Geopolymerisation of multiple minerals. Miner Eng 15:1131–1139
26. Sears GW (1956) Determination of specific surface area of colloidal silica by titration with sodium hydroxide. Anal Chem 28:1981–1983

Chapter 21
Prediction of Long-Term Chemical Evolution of a Low-pH Cement Designed for Underground Radioactive Waste Repositories

T. Bach, I. Pochard, C. Cau-dit-Coumes, C. Mercier, and A. Nonat

1 Introduction

A solution to manage long-lived radioactive wastes over the very long time is their disposal in deep geological structures. The disposal concepts are based on a multibarrier design approach including the waste package (the waste and the material used to stabilize it in a suitable overpack), the engineered barrier inserted between the waste package and the rock and the geological barrier, i.e. the actual rock. In most designs, bentonite is a key component of the engineered barrier system. It is however unstable at high pHs (≈12–13) [1], such as those encountered in ordinary concrete leachates. Since it seems more and more difficult to exclude concrete from any sensitive areas containing bentonite in a repository, some work has been undertaken in order to develop low-pH cements which would show an improved compatibility with clay. Low-pH cements, also referred as low-alkalinity cements, are binders with a pore solution pH $\leq$ 11. They can be designed by replacing significant amounts of Portland cement (OPC) ($\geq$40 %) by silica fume, which can be associated in some cases to low-CaO fly ash and/or ground granulated blast furnace slag [2–4]. Such a blend has several positive effects: (1) consumption

T. Bach
Commissariat à l'Energie Atomique et aux Energies Alternatives, DEN/MAR/DTCD/SPDE, BP17171, 30207 Bagnols-sur-Cèze, France

ICB, UMR 5209 CNRS Université de Bourgogne, 21078 Dijon, France

I. Pochard (✉) • A. Nonat
ICB, UMR 5209 CNRS Université de Bourgogne, 21078 Dijon, France
e-mail: isabelle.pochard@u-bourgogne.fr

C. Cau-dit-Coumes
Commissariat à l'Energie Atomique et aux Energies Alternatives, DEN/MAR/DTCD/SPDE, BP17171, 30207 Bagnols-sur-Cèze, France

C. Mercier
LMCPA, Université de Valenciennes et du Hainaut Cambrésis, 59313 Valenciennes cedex, France

F. Bart et al. (eds.), *Cement-Based Materials for Nuclear Waste Storage*,
DOI 10.1007/978-1-4614-3445-0_21,

of portlandite, which is formed by the hydration of OPC and which buffers the pore solution pH up to 12.5, to produce calcium hydrosilicates (C-S-H) by the pozzolanic reaction, (2) decrease in the CaO/SiO_2 ratio of the C-S-H and partial substitution of Al^{3+} for Si^{4+} in the silicate chains of the C-S-H, which could enhance their sorption capacity of alkalis [5], and (3) dilution of OPC, which decreases the heat output during hydration. The pore solution pH of the cement paste is then controlled by the solubility of C-S-H: the smaller the CaO/SiO_2 ratio of C-S-H, the lower the pH [6, 7]. With the prospect of using these materials in a geological repository, it is of main importance to estimate their long-term properties and the influence of external and internal factors (chemical composition of the binder, storage temperature) on their characteristics at equilibrium. This chapter describes the mineralogical evolution of a low-pH binder comprising OPC, silica fume, and fly ash over one year. Since hydration was still incomplete after this period, a three-way original approach was adopted to assess the behaviour of this hydrated low-pH cement at thermodynamic equilibrium, i.e. at the end of its evolution under fixed conditions in the temperature range 20–80°C. Firstly, thermodynamic modelling was carried out to predict the mineral assemblage and composition of the solution at equilibrium, starting from the composition of the initial low-pH cement studied. Secondly, hydration of blended cements was accelerated by milling cement slurries (water to solid ratio of 9) with zirconia beads. Thirdly, the low-pH cement pastes were mimicked with mixtures of appropriate highly reactive oxides (lime, silica, calcium aluminate, and calcium sulphate) with a water to solid ratio of 50.

2 Materials and Methods

2.1 Raw Materials and Cement Pastes

The low-pH binder investigated in this work comprised 37.5 % Ordinary Portland Cement, 32.5 % Silica Fume, and 30 % Fly Ash. Table 21.1 gives the chemical composition and properties of the raw materials used in this formulation. Note that the quite important proportion of "minors" in the fly ash is due to a significant loss on ignition (near 5 %). The binder constituents were ground together in a planetary ball mill (Retsch PM400 mixer-mill: 250 rpm, 150 g of material, and 50 g of 10 mm diameter stainless steel balls in a 250 mL grinding jar). Reference cement pastes were prepared at water to solid ratio of 0.55, stored in hermetically sealed 50 mL polypropylene containers, and submitted to three types of curing (1) in a controlled-environment chamber at 20°C and 95 % relative humidity, in an oven at (2) 50°C, or (3) 80°C. To prevent drying, the samples were placed in water-filled boxes. Curing was initiated 30 min after mixing. All the binders were characterized after 3, 6, and 12 months of hydration. A detailed procedure for the solid and pore solution characterization is presented elsewhere [8].

Table 21.1 Characteristics of binder constituents

		CEM I	Silica fume	Fly-ash
Chemical composition (wt%)	CaO	66.90	0.50	5.10
	SiO_2	22.00	96.30	51.50
	Al_2O_3	3.30	<0.20	25.20
	Fe_2O_3	2.74	0.10	5.80
	MgO	0.60	<0.20	1.80
	Sulphates	2.40	0.24	0.66
	$Na_2O + K_2O$	<0.25	<0.54	1.73
	Minors	1.81	1.92	8.21
Phases observed by X-ray diffraction[a]		C_3S^{+++} C_2S^{+++} Brownmillerite $^{+}$ C_3A^{+} Gypsum $^{+}$ Anhydrite $^{+}$ Calcite $^{+}$	Cristobalite $^{+++}$ Quartz $^{++}$ Maghemite $^{++}$ Silicon $^{+}$ Silicon oxide $^{+}$	Quartz $^{+++}$ Mullite $^{++}$ CaO $^{+}$ Maghemite $^{+}$

[a]Intensity of main diffraction peak: $^{+++}$ strong, $^{++}$ moderate, $^{+}$ weak

2.2 *Acceleration of the Hydration by Milling*

To accelerate the reaction, slurries of low-pH cement were submitted to milling instead of simple curing. By means of the stirring, the zirconia beads ground away the hydrates formed on the grain surfaces and thus re-exposed the anhydrous phases to water. The mechanical energy input to the system is able to change the nature of the hydrosilicate phase. For instance, the same procedure for the hydration of C_3S with aluminium balls leads to the formation of afwillite [9]. In the present study, this phase was never observed. The tests were conducted at controlled temperature (20, 50, or 80°C) in the absence of carbon dioxide. The water/binder ratio was set at 9 mL/g by the addition of 5 g of binder to 45 mL of deaerated deionized water. For curing at 20°C, 51 g of zirconia beads (diameter: 0.8 to 2.5 mm) were added to the slurry in hermetically sealed polypropylene containers continuously stirred in a turbula for 3 months. The samples at 50 and 80°C were milled with 30 g of zirconia beads (diameter: 2–2.5 mm) in hermetically sealed Teflon containers under continuous magnetic stirring in an oven for 2 months at the selected temperature. To prevent drying, the samples were placed in water-filled boxes. After this equilibration time, the suspensions were allowed to stand without stirring at the studied temperature for 1 month. The pH was determined at the sample curing temperature (20, 50, or 80°C). The high alkalinity measurement electrode was first calibrated at the same temperature with buffer solutions: borate (pH 9.23 at 20°C, 9.01 at 50°C, and 8.88 at 80°C), carbonate (pH 10.06 at 20°C, 9.83 at 50°C, and 9.73 at 80°C), and calcium hydroxide (pH 12.63 at 20°C, 11.71 at 50°C, and 10.99 at 80°C). The suspension was then filtered at the test temperature and the filtrate was acidified. Its composition was analysed by ICP-AES (Vista Pro Varian, Ca, Si, Al, S, Na, K standardization with matrix reconstitution). The solids were recovered by using successive water/ethanol and pure ethanol washing/drying. The mineralogy of the cement pastes was characterized by X-ray diffraction (Siemens D8, copper anode, $\lambda_{K\alpha 1} = 1.54056$ Å, 40 mA and 40 kV, scanning from $2\theta = 5°$ to $60°$ in $0.017°$ steps, 50 s measurement time per step).

2.3 Reactive Oxides in Suspension

The tests were conducted at controlled temperature in the absence of carbon dioxide on dilute suspensions (water/solid ratio = 250 mL/5 g). The hydrated binders were precipitated from mixtures of 1.05 g of CaO ($CaCO_3$ Prolabo decarbonated at 1,000°C for 3 h), 2.75 g of SiO_2 (Aerosil 200), 0.69 g of calcium aluminate ($CaAl_2O_4$) and 0.13 g of calcium sulphate ($CaSO_4$, $2H_2O$), and 250 g of 7 mmol/L alkaline aqueous solution. These quantities are so that the Ca, Si, Al, SO_4, and alkali proportions are the same as in the low-pH binder (Ca/Si = 0.52; Al/Si = 0.19). These reagents were chosen for their high reactivity in water. Compared to the actual composition of the ternary binder, the composition of the model systems was simplified. Indeed, the minor elements (Fe, Mg, Mn, Ti, P) were not taken into account. The suspensions were stored in hermetically sealed 250 mL polypropylene containers under continuous stirring for 2 months at 20°C, 3 weeks at 50°C, and 1 week at 80°C. The solid and aqueous fractions were characterized in the same way as the accelerated hydration samples.

2.4 Thermodynamic Calculations

Thermodynamic modelling was carried out to predict the mineral assemblage and composition of the solution at equilibrium, starting from the composition of the initial low-pH cement studied. Simulations were performed with PHREEQC software provided by US Geological Survey [http://www.USGS.gov] and using the database developed by Damidot et al. for cement-based systems [10–14]. Simulations were ruled by the solubility products which governed the actual system. The goal of these calculations was twofold. It could point out the possible presence of amorphous phases not observable by XRD. Besides, comparing the calculated results with phase assemblages obtained by the two experimental methods could help predicting the final state of equilibrium of the fully hydrated low-pH cement paste. The minor elements were not taken into account, and the crystalline phases from fly ash were treated as the amorphous reactive part. This means that it is the total amounts of Ca (8.69 mol/kg), Si (16.67 mol/kg), and Al (3.16 mol/kg) present in the low-pH cement formulation, including all phases from fly ash that were the input data in mol per kg of water. Taking into account the solubility of such phases was not necessary in a first approach because their amount was negligible compared to the large proportion of amorphous silica and alumina present in silica fume and fly ash. The supersaturated phases were let to precipitate and the ionic solution was charge balanced with pH. Sulphates and alkalis adsorptions on C-S-H were taken into account to calculate their concentrations at equilibrium [15, 16]. The substitution of Al^{3+} for Si^{4+} in the C-S-H was not taken into account. For a more accurate description, it should not be neglected. This will be the subject of future work.

3 Results and Discussion

3.1 Low-pH Cement Pastes over 6 months

The mineralogy of the reference cement pastes was investigated at 20, 50, and 80°C. Figure 21.1 illustrates the XRD results recorded on a 6-month-old sample. The pastes mainly consisted of calcium silicate hydrate, probably containing aluminium (C-A-S-H). No portlandite was detected, as expected for low-pH cements. The disappearance of ettringite was observed above 50°C. It is well known that the stability range of ettringite decreases in favour of calcium monosulfoaluminate hydrate when temperature increases [17]. The absence of ettringite for samples cured above 50°C ensures that the procedure used to recover the solids after curing is suitable and does not lead to mineralogy change during filtration and drying. Obviously, the crystalline phases of the fly ash (mullite, quartz, hematite), which are unreactive, were still detected. Although C_3S was not observed, the presence of C_2S showed that hydration was not complete after

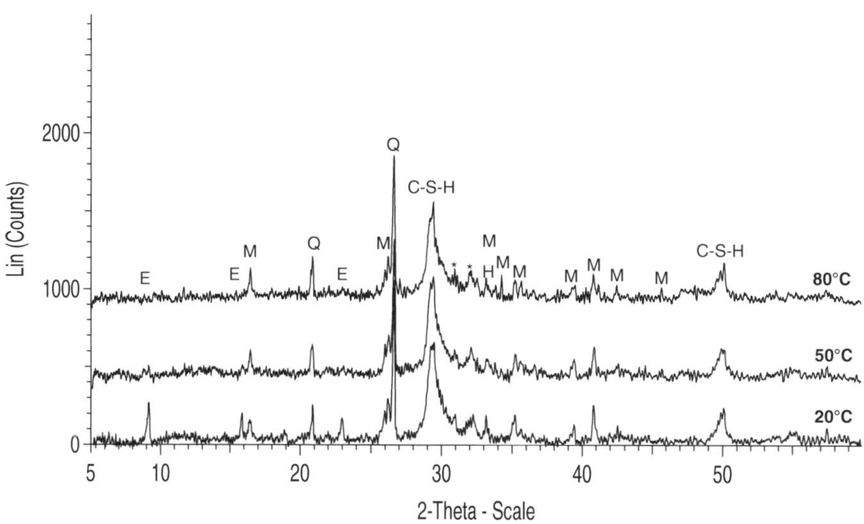

Fig. 21.1 XRD patterns of the low-pH cement pastes cured 6 months at 20, 50, and 80 °C. *E* ettringite, *M* mullite, *Q* quartz, *H* hematite, *asterisk* C_2S

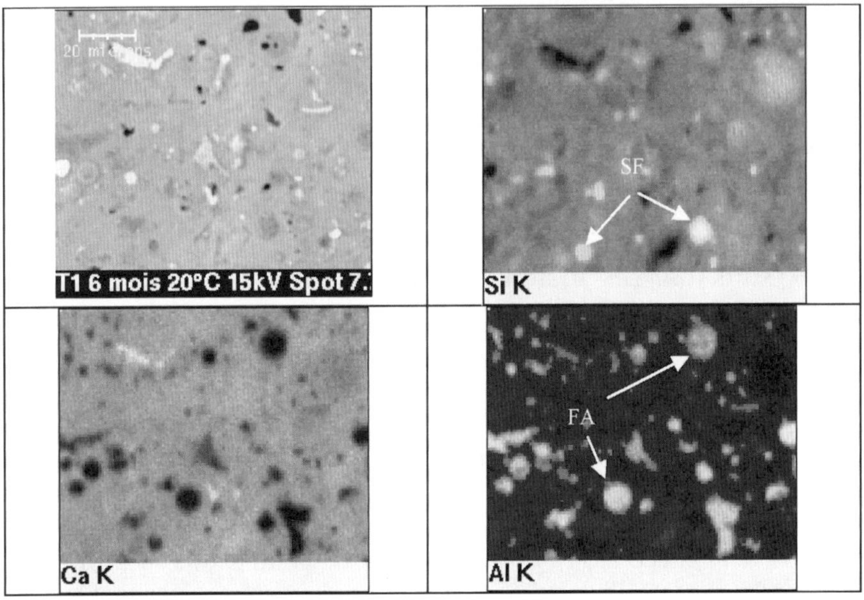

Fig. 21.2 Elemental analysis mapping of the low-pH cement paste after 6 months of curing at 20 °C. *SF* residual silica fume, *FA* residual fly ash

Table 21.2 Composition (in mmol/L) and pH of the liquid fraction of the cement pastes after 6 months curing

T(°C)	Ca	Si	S	Al	Alkali	pH
20	3.2	2.1	3.2	<d.l.	3.0	11.0
50	6.0	2.0	6.6	0.01	3.6	10.0
80	6.5	3.0	7.4	<d.l.	4.2	9.1

<d.l. means below the detection limit

this time. Indeed, hydration was very slow after 3 months and remained incomplete even after 1 year.

The EDS(SEM) mapping was used in addition to XRD to determine the mineralogy of the pastes as this technique enables the detection of amorphous compounds. Figure 21.2 shows that there were still amorphous silica fume and fly ash in the low-pH cement paste at 20°C after 6 months of hydration. This was not surprising as these two constituents were present in large proportions in the formulation.

Finally, the description of the pore solution after 6 months and for the three temperatures is given in Table 21.2.

The pH reduction mainly resulted from the temperature-dependent variation of the activity coefficients [18]. Raising the curing temperature resulted in an increase in the sulphate concentration in solution. This was due to the disappearance of ettringite from the samples stored at 50°C and 80°C, without any significant

Table 21.3 Thermodynamic simulation results of the low-pH cement paste fully hydrated at 20 °C

	Initial quantity[a] (mol/kg water)	Calculated concentration[b] (mM)	pH	Solid phases[c]
Ca	8.69	1.79	10.20	C-S-H, silica, $Al(OH)_3$
Si	16.67	5.06		
Al	3.16	0.05		
SO_4	0.22	1.00		
Alkalis	0.36	2.31		

[a]Based on the composition of the low-pH binder
[b]Concentrations in the solution in equilibrium with
[c]Precipitated solid phases

precipitation of other sulphate phases. The concentrations of the other species tended to increase more slowly with the temperature, which could mainly result from the variation of the activity coefficients [18].

3.2 Determination of the Final State by Thermodynamic Calculation

Even if the simulated system was simplified compared to the actual low-pH cement paste, several conclusions could be drawn. The results at 20°C are listed in Table 21.3. Compared to the cement paste at 20°C, there was no C_2S remaining, but silica and $Al(OH)_3$ were still present besides C-S-H at the final stage. In the cement paste, residual silica was also noticed as amorphous silica and quartz from silica fume, and as amorphous and crystalline phases from fly ash. At equilibrium, no free Al was available, which was already the case in the cement paste cured for 6 months. Actually, aluminium was precipitated in $Al(OH)_3$ and/or unreacted fly ash. The Ca and Si concentrations fitted the C-S-H solubility curve for a Ca/Si ratio equal to 0.66. On the contrary, the Ca concentration in the pore solution of the 6-month-old cement paste was higher (3.2 mM at 20°C), which was characteristic of a higher Ca/Si ratio. The pozzolonic reaction could thus continue after 6 months of curing. This was corroborated by the calculated pH value, almost one unit smaller than that measured on the cement paste.

The final state of the low-pH formulation at 20°C was thus composed of C-S-H of Ca/Si = 0.66 with excess of silica and $Al(OH)_3$ coming from the excess of silica fume and fly ash. The C-S-H may include aluminium by Si/Al substitution but this feature could not be taken into account by the simulation. However, thc Al content in the C-S-H should be low given the very low Al concentration in the pore solution [19]. The pH of the pore solution was thus governed by the C-S-H solubility. Ettringite, which precipitated at early age when hydration occurred at 20°C, was shown to be unstable at thermodynamic equilibrium. Moreover, apart from the destabilization of ettringite and the variation of the activity coefficients [18], this

Table 21.4 Crystalline phases in the solids, composition (in mmol/L) and pH of the liquid fraction of the cement suspensions obtained by milling, and of the mixed oxides suspensions at 20, 50, 80°C

	20°C		50°C		80°C	
	Hydration by milling	Model systems	Hydration by milling	Model systems	Hydration by milling	Model systems
Crystalline phases[a]	C-S-H +++ Mullite ++ Quartz ++ Calcite +	C-S-H +++ Ettringite +	C-S-H +++ Mullite ++ Quartz ++ Calcite +	C-S-H	C-S-H +++ Mullite ++ Quartz ++ Calcite +	C-S-H
Ca	5.5	1.2	10.0	2.2	8.8	2.8
Si	1.9	3.7	2.2	3.3	1.8	2.6
Al	<d.l.	<d.l.	<d.l.	<d.l.	<d.l.	<d.l.
SO_4	8.2	1.0	13.5	2.4	12.3	3.1
Alkali	9.4	2.3	9.3	2.4	8.8	2.1
pH	10.7	10.4	9.7	9.5	9.0	9.2

<d.l. means below the detection limit
[a]Intensity of main diffraction peak: +++ strong, ++ moderate, + weak

equilibrium state should not strikingly change with temperature. The pozzolonic reaction was indeed already complete at 20°C (there was no more calcium hydroxide reservoir and the C-S-H got its lower Ca/Si ratio). Besides, the well-known increase in silicate chain polymerization of the C-S-H with temperature [20] should not occur as the C-S-H already gets the lowest Ca/Si ratio at 20°C.

3.3 *Hydration by Milling and Reactive Oxides in Suspension*

The goal of these approaches was to estimate if it was possible to reach equilibrium experimentally by accelerating the reactions. The crystalline phases present in the solid fractions and the ionic composition of the aqueous fractions of the samples obtained by milling and by mixing oxides in dilute suspensions are listed in Table 21.4 at the three temperatures.

All the samples contained C-A-S-H, but were free from portlandite and C_2S. Traces of ettringite were sometimes observed in model samples at 20°C. Using the milling technique, the unreactive phases of fly ash (mullite, quartz) remained present. The samples were not systematically analysed by EDS. However, the few analyses done showed the presence of non-reacted amorphous silica and $Al(OH)_3$ in both systems (milling and oxides in suspension). It corroborates the fact that the silica fume and fly ash were present in excess compared to OPC in the low-pH formulation. The mineral assemblages were thus similar in the three cases: hydration by milling, oxides in dilute suspensions, and thermodynamic simulation. Nevertheless, the phase proportions may vary from one case to another depending on the pozzolanic reaction progress. As a matter of fact, the Ca/Si of the C-A-S-H would decrease as well. In the cement pastes, the only supplementary phase was C_2S.

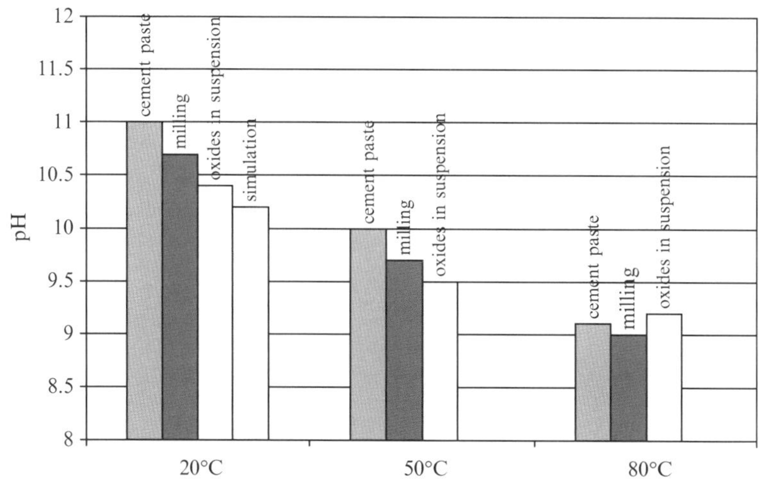

Fig. 21.3 Comparison of the pH values obtained with the different methods at the three studied temperatures

At 20°C, the pH decreased within the series cement paste/milling/oxides in suspension/thermodynamic simulation (see Fig. 21.3). This pH decrease was correlated with a Ca concentration decrease and Si concentration increase, except for the milling experiment (in this case, the sulphate concentration was surprisingly high, involving a high Ca concentration as well). These variations may thus be related to a greater progress of the pozzolanic reaction. The same pH decrease was observed at 50°C and 80°C between the three experimental methods, but the deviations were less and less significant when the temperature rose. At 80°C, the pH values were similar considering the experimental errors. It tends to show that the increase in temperature accelerated the pozzolanic reaction and at 80°C, the reaction progress was similar in the three experiments.

In conclusion, the two experimental approaches made it possible to consume all the reactive anhydrous phases from the low-pH binder. The small differences observed at 20°C and 50°C compared to thermodynamic simulation could be explained by a slightly less advanced pozzolanic reaction.

4 Conclusion

With the prospect of using low-pH binders in a geological repository, the goal of this work was to estimate the long-term chemical evolution of such materials. For this purpose, a three-way original approach was adopted, which consisted in the comparison of the final state achieved by thermodynamic simulations with experimental data. These data were obtained by milling the cement paste or by mixing reactive oxides in suspension. Comparing the different results showed that this three-way approach is suitable to understand and predict the long-term chemical evolution of cements since the final states obtained in all cases were equivalent.

The mineral assemblages were similar in the three cases. At 20°C and 50°C, the phase proportions could slightly vary from one case to another, depending on the progress of the pozzolanic reaction. Raising the temperature from 20°C to 80°C accelerated the pozzolanic reaction and, at 80°C, similar results were obtained whatever the experimental method. In conclusion, the three-way approach under investigation was suitable to assess the chemical properties of fully hydrated cementitious materials with complex chemistry.

References

1. Huertas FJ (2005) Effect of cement on clay barrier performance - phase II, European Commission, EUR21921 report. ECOCLAY II – FIKW – CT – 2000 – 00018
2. Cau-dit-Coumes C, Courtois S, Nectoux D, Leclercq S, Bourbon X (2006) Formulating a low alkalinity, high resistance and low heat concrete for radioactive waste repositories. Cement Concr Res 36:2152–2163
3. Codina M, Cau-dit-Coumes C, Le Bescop P, Verdier J, Ollivier JP (2008) Design and characterization of low-heat and low-alkalinity cements. Cement Concr Res 38:437–444
4. Garcia Calvo JL, Hidalgo A, Alonso C, Fernandez-Luco L (2010) Development of low-pH cementitious materials for HLRW repositories. Cement Concr Res 40:1290–1297
5. Hong S, Glasser FP (2002) Alkali sorption by C-S-H and C-A-S-H gels: Part II. Role of alumina. Cement Concr Res 32:1101–1111
6. Flints EP, Wells LS (1934) Study of the system CaO-SiO_2-H_2O at 30°C and of the reaction of water on the anhydrous calcium silicates. Bur Stand J Res 12:751–783
7. Roller PS, Erwin G (1940) The system calcium oxyde silica water at 30°C - The association of silicate ion in dilute alkalin solution. J Am Chem Soc 62:461–471
8. Bach TTH, Cau-dit-Coumes C, Pochard I, Mercier C, Nonat A, Influence of temperature on the hydration products of low pH cements. Submitted to CCR
9. Kantro DL, Brunauer S, Weise CH (1959) The ball-mill hydration of tricalcium silicate at room temperature. J Colloid Int Sci 14:363–376
10. Damidot D, Glasser FP (1993) Thermodynamic investigation of the CaO-Al_2O_3-$CaSO_4$-K_2O-H_2O at 25°C. Cement Concr Res 23:1195–1204
11. Damidot D, Birmin-Yauri U, Glasser FP (1994) Thermodynamic investigation of the CaO-Al_2O_3-$CaCl_2$-H_2O system at 25°C and the influence of Na_2O. Cemento 91(4):243–254
12. Damidot D, Stronach S, Kindness A, Atkins M, Glasser FP (1994) Thermodynamic investigation of the CaO-Al_2O_3-$CaCO_3$-H_2O closed system at 25°C and the influence of Na_2O. Cement Concr Res 24(3):563–572
13. Damidot D, Nonat A, Barret P, Bertrandie D, Zanni H, Rassem R (1995) C3S hydration in diluted and stirred suspensions: (III) NMR study of C-S-H precipitated during the two kinetic steps. Adv Cement Res 7(25):1–8
14. Damidot D, Glasser FP (1995) Investigation of the CaO-Al_2O_3-SiO_2-H_2O system at 25°C by thermodynamic calculations. Cement Concr Res 25(1):22–28
15. Barbarulo R, Peycelon H, Leclercq S (2007) Chemical equilibria between C-S-H and ettringite, at 20 and 85°C. Cement Concr Res 37:1176–1181
16. Bach TTH (2010) Evolution physico-chimique des liants bas pH hydratés : influence de la température et mécanisme de rétention des alcalins. PhD thesis, University of Bourgogne

17. Damidot D, Glasser FP (1992) Thermodynamic investigation of the CaO-Al_2O_3-$CaSO_4$-H_2O system at 50°C and 85°C. Cement Concr Res 22:1179–1191
18. Dean JA (1999) Lange's handbook of chemistry, 15th edn. Mc Graw Hill, New York
19. Pardal X, Pochard I, Nonat A (2009) Experimental study of Si-Al substitution in calcium-silicate-hydrate (C-S-H) prepared under equilibrium conditions. Cement Concr Res 39:637–643
20. Young JF (1988) Investigations of calcium silicate hydrate structure using 29Si nuclear magnetic resonance spectroscopy. J Am Ceram Soc 71(3):118–120

Chapter 22
Round Robin Test for Defining an Accurate Protocol to Measure the Pore Fluid pH of Low-pH Cementitious Materials

M.C. Alonso, J.L. García Calvo, S. Pettersson, I. Puigdomenech, M.A. Cuñado, M. Vuorio, H. Weber, H. Ueda, M. Naito, C. Walker, Y. Takeshi, and C. Cau-dit-Coumes

1 Introduction

The production of low pH blends cements to fit with the engineering barrier concept requirements to be used in High Radioactive Waste Repositories (HLWR) must be based on the physicochemical requirements of the cementitious material developed

M.C. Alonso (✉) • J.L.G. Calvo
Research Centre in Safety and Durability of structures and materials, CISDEM (CSIC-UPM), Madrid, Spain
e-mail: mcalonso@cisdem.upm-csic.es

S. Pettersson • I. Puigdomenech
Swedish Nuclear fuel and waste management, SKB, Stockholm, Sweden

M.A. Cuñado
Empresa Nacional de Residuos Radioactivos, ENRESA, Madrid, Spain

M. Vuorio
Nuclear waste Management in Finland, Posiva, Eurajoki, Finland

H. Weber
National Cooperative for the Disposal of Radioactive Waste in Switzerland, NAGRA, Wettingen, Switzerland

H. Ueda
Nuclear Waste Management Organization of Japan, NUMO, Tokyo, Japan

M. Naito • C. Walker
Japan Atomic Energy Agency in Japan, JAEA, Aomori, Japan

Y. Takeshi
Central Research Institute of Electric Power in Japan, CREIPI, Tokyo, Japan

C. Cau-dit-Coumes
Commissariat à l'Energie Atomique et aux Energies Alternatives, DEN/MAR/DTCD/SPDE, Bagnols-sur-Cèze, France

F. Bart et al. (eds.), *Cement-Based Materials for Nuclear Waste Storage*,
DOI 10.1007/978-1-4614-3445-0_22,

(fluid properties, mechanical generation properties, permeability properties, pH requirement, etc.). Hence, an accurate measurement of the pore fluid pH is needed in order to qualify mix designs and to select low-pH cements for repository use. The methods more often used in the literature for pore solution characterization (including pH) of cementitious materials are Pore Fluid Expression (PFE) and Extraction Leaching methods (ESL), including both in situ and ex situ techniques.

The basic concept of the Pore Fluid Expression (PFE) method is to extract the pore fluid of hardened cementitious materials under high pressure[1–3] and then analyze its chemical composition. The main limitations of the PFE method are the requirement of a specialized and relatively expensive press and of quite a large amount of sample, as well as being quite time consuming to perform. Ex situ Leaching (ESL) methods rely on a sample of "pore fluid" obtained by mixing a known amount of crushed or powdered cementitious material with a known amount of deionized water [4–13].

However, in spite of the published experiences or the different method types used, as PFE or ESL, there are no agreed protocols or standards to use for pore fluid pH characterization of cementitious materials. In the present chapter, eight laboratories were involved in Round Robin tests in order to evaluate the repeatability and reproducibility of the PFE and ELS methods developed in four different types of low-pH cementitious (LopHC) materials.

2 Experimental

Four LopHC materials were prepared in different laboratories and Table 22.1 presents their compositions. The composition of CM 1 and CM 2 are given in kg/ m^3 while CM 3 and CM 4 are expressed in g per 100 g of binder. Samples of each LopHC material were distributed amongst the eight laboratories involved in the Round Robin test. The pore fluid pH of every LopHC material was measured by each laboratory following identical protocols.

Table 22.1 Composition of the four LopHC materials for pH characterization

Material	CM 1 (kg/m^3)	CM 2 (kg/m^3)	CM 3 (g per 100 g binder)	CM 4 (g per 100 g binder)
OPC	120	165	50	60
Silica fume	80	110	50	–
Fly Ash	–	–	–	40
Water	165	230	125	35
Limestone filler	369	70	–	–
Sand 0–8 mm	1 037	1066	–	–
Gravel 8–16 mm	558	593	–	–
Superplasticizer	6.38	5.1	5.5	–
Water/cement	1.375	1.39	2.5	0.58
Water/binder	0.825	0.84	1.25	0.35
Water/powder	0.29	–	–	–

Each LopHC material was cured in a 100 % Relative Humidity (RH) chamber at 21 ± 2 °C before being shipped to the eight different laboratories in sealed plastic bags to prevent CO_2 contamination and drying of the samples during transport. The samples received in each laboratory were stored in a 100 % RH chamber at 21 ± 2 °C until testing after a total of 90 days curing ±5 days.

The tests used were:

1. PFE performed in two laboratories and applied to samples CM 3 and CM 4. The pH values of CM 1 and CM 2 concrete samples were not measured using PFE due to the difficulty and risk of damage of the pore solution expression equipment generated by the coarse aggregates.
2. ESL in the 4 LopHC, with (a) filtering the suspension and (b) no-filtering, both used in the eight laboratories.

   Considering all the pH measurements made, 12 analyses were done using the PFE method and 54 using each variant of the ESL method.

In all cases, the pH was determined by direct measurement using a pH meter with ability of temperature compensation and a combination pH electrode specific to measuring in the pH range 7–14. Besides, in the expressed solutions (using PFE method) and in the filtered one (using ESL method), the pH was determined by OH^- titration and further calculation of the pH. A noncellulose filter should be used for filtering the suspension.

2.1 Pore Fluid Expression Protocol

Approximately 125 g of cementitious (LopHC) material was divided into pieces of $\phi < 0.5$ cm. As soon as possible, the mixture was placed in the cylinder of the pore pressing device. Pressure was increased in steps of approximately 50 MPa/min, until a maximum load pressure of 483 MPa (70 Psi) was achieved, and then maintained for 5 min to express a sample of the pore solution, Fig. 22.1a. The pore solution sample was collected in a syringe inserted into the hole of the bottom plate of the press (Fig. 22.1b), transferred to a controlled N_2 atmosphere (bubbling N_2 gas), passed through a 0.45 μm filter (preferable no cellulose based) and stored in a sealed plastic container until analysis, Fig. 22.1c and 22.1d. Direct pH measurements, under a controlled N_2 atmosphere, must be done as soon as possible after the pore fluid expression. Three different samples were used to determine the pH of the pore fluid from each laboratory and material.

2.2 Ex Situ Leaching Protocol

A sample of each material was finely ground to a powder particle size (around 80 μm). Three powdered portions of 10 g were each mixed in a beaker with 10 mL of CO_2-free deionized water (solid/liquid = 1.0) and the suspension continuously and

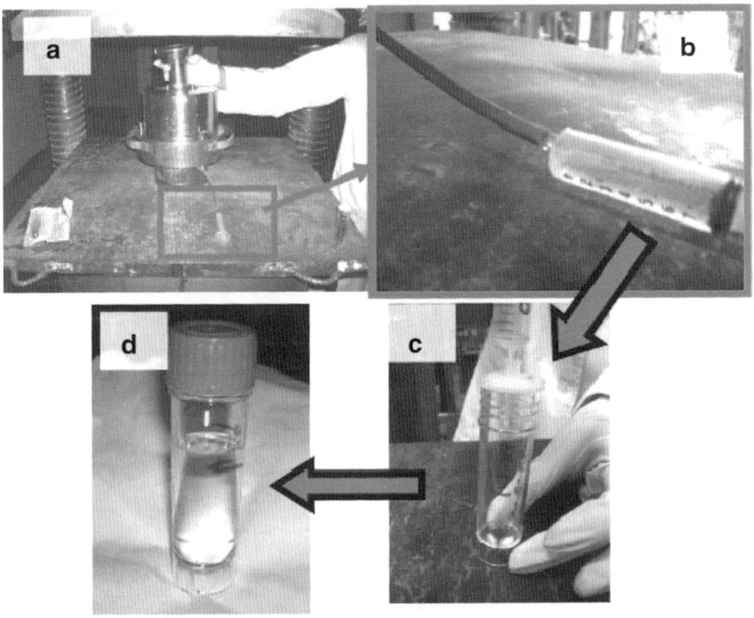

Fig. 22.1 Equipment and procedure for PFE method

Fig. 22.2 Procedure for pH ESL without filtering (*left*) and filtering (*right*)

vigorously stirred for 5 min, Fig. 22.2, left. After stirring, two procedures were followed: (1) to measure the pH of the suspension directly with a pH electrode under a N_2 atmosphere, stirring continuously during the measurement, Fig. 22.2, centre; (2) to pass the suspension through a 0.45 μm filter and measure the pH of the filtrate (a noncellulose-based filter is preferable to avoid possible influences in the solution), Fig. 22.2, right. The stirring and both measurement procedures were done under a N_2 atmosphere (by passing N_2 gas across the top of the suspension or the solution surface). The procedure was repeated nine times with each of the 4 LopHC samples, so that nine identical measurements were taken in each laboratory for each material.

3 Results and Discussion

Table 22.2 presents the mean pH values obtained for each cementitious material. Except when the pH is measured by OH^- titration using the ESL method, similar results are obtained by the different methods used for each cementitious material after 90 ± 5 days of curing. Note that, in general, the Standard Deviation (SD) is close to or within the ±0.1 error that can be assigned to the pH electrode. Moreover, the methods used allow a rapid test and classification/definition of a candidate LopHC material (pH ≤11). The high pH value of the CM 4 LopHC paste is related to the use of FA as an additive, which is well known to have a very slow pozzolanic reaction on the order of years.

3.1 *Statistical Analysis of the Repeatability and the Reproducibility of Both ESL Methods*

The number of measurements performed using the ESL protocols has allowed to carry out statistical analysis and to determine the Repeatability r and Reproducibility R of the ESL methods, with and without filtering. The analysis was performed following the suggestions of the method proposed in the ISO 5725 standard [14]. The procedure was similar to that described in [15]. Table 22.3 shows the statistical parameters obtained.

The main equations used, according to ISO 5725, in the statistical study of the repeatability and the reproducibility of the ELS methods are given by:

For the estimation of the repeatability variance (S_r^2):

$$S_r^2 = \frac{SS_r}{N - p} \tag{22.1}$$

where

SS_r = residual or intra-laboratory sum of squares,
p = number of laboratories,
N = total number of data.

Table 22.2 Summary of mean pH values and ± SD obtained from the PFE and ESL methods of the 4 LopHC materials

Cementitious material	PFE		ESL		
			Non-filtering	Filtering	
	pH electrode	OH^- titration	pH electrode	pH electrode	OH^- titration
CM 1	–	–	11.25 ± 0.14	11.14 ± 0.15	11.27 ± 0.26
CM 2	–	–	10.95 ± 0.11	10.81 ± 0.09	11.09 ± 0.26
CM 3	10.71 ± 0.13	10.74 ± 0.15	11.13 ± 0.10	10.80 ± 0.30	11.16 ± 0.46
CM 4	13.01 ± 0.01	13.02 ± 0.21	12.77 ± 0.05	12.73 ± 0.06	12.43 ± 0.44

Table 22.3 Statistical parameters for the ESL methods without/with filtering the suspension

	LopHC materials							
	CM 1		CM 2		CM 3		CM 4	
S_r	0.04	0.05	0.042	0.05	0.04	0.02	0.03	0.01
r	**0.11**	**0.14**	**0.12**	**0.15**	**0.12**	**0.07**	**0.09**	**0.04**
V_r/%	0.36	0.46	0.38	0.51	0.37	0.23	0.26	0.11
S_R	0.14	0.14	0.12	0.09	0.08	0.11	0.06	0.06
R	**0.40**	**0.41**	**0.35**	**0.27**	**0.23**	**0.32**	**0.16**	**0.18**
V_R/%	1.25	1.30	1.13	0.87	0.72	1.02	0.45	0.50

For value of repeatability:

$$r = 2.83 S_r \tag{22.2}$$

For percentage of variation or repeatability variability (V_r):

$$V_r = 100\, S_r / m \tag{22.3}$$

where m = gross average value.

For estimation of inter-laboratory variance ($S_L{}^2$):

$$S_L^2 = \frac{(p-1)\left(\frac{SS_L}{p-1}\right) - S_r^2}{N'} \tag{22.4}$$

where

SS_L = factorial or inter-laboratory sum of squares,
N' = corrected total number of data.

For estimation of reproducibility variance ($S_R{}^2$)

$$S_R^2 = S_L^2 + S_r^2 \tag{22.5}$$

For value of reproducibility

$$R = 2.83 S_R \tag{22.6}$$

For percentage of variation or reproducibility variability (V_R):

$$V_R = 100\, S_R / m \tag{22.7}$$

where m = gross average value.

By analyzing the statistical parameters, several conclusions can be drawn: (1) Both measurement procedures show similar repeatability (r) and reproducibility (R) values, so they can be considered as reliable. (2) The repeatability of both ESL methods is very good, as the so calculated r values are quite close or even lower than the ±0.1 error that can be assigned to the pH electrode. The calculated r values imply that, in the worst case, with a probability of 95 %, the difference between two

Table 22.4 w ratios corresponding to the ESL methods

	Samples	
Method	CM 3 (w)	CM 4 (w)
ESL non-filtering	1.78	1.71
ESL filtering	0.53	1.93

different individual results obtained under the same conditions will be below 0.15 pH units, which indicates that the developed protocols are quite accurate. A pH unit difference of 0.15 also represents less than 0.5 % of the pH values measured.

Calculated values of r are significantly better (i.e., lower) for the LopHC pastes (CM 3 and CM 4) than for the LopHC concretes (CM 1 and CM 2) in using the ESL method with filtering. This can be attributed to a paste being more homogeneous than concrete.

In summary, this statistical study demonstrates that the developed routine (ESL) methods show satisfactory repeatability and reproducibility values when they are used for different LopHC materials, even with high alkaline pH values (CM 4).

3.2 Comparison of the ESL Methods to the PFE Method: Statistical Analysis

A critical evaluation of the accuracy of the ESL methods with respect to the measured pH values using the PFE technique can only be made for the two LopHC paste materials (CM 3 and CM 4) according to the ISO 5725 standard [14], performing the trueness comparison in the same way as that described in [15]. The calculation of the w ratio has been made considering both Routine ESL Methods for each LopHC paste material (CM 3 and CM 4) individually; the results are shown in Table 22.4, for which the outliers have been excluded.

According to the ISO 5725 standard, the trueness comparison is performed by computing the ratio w in the following way:

$$w = \frac{\left|\bar{\bar{x}}_{\mathrm{M}} - \bar{\bar{x}}_{\mathrm{A}}\right|}{\delta} \tag{22.8}$$

where

$\bar{\bar{x}}_{\mathrm{M}}$= gross average of the Reference PFE Method,
$\bar{\bar{x}}_{\mathrm{A}}$= gross average of the Routine ESL Methods,

and δ is given by:

$$\delta = 2\sqrt{\frac{s_{\mathrm{LA}}^2 + s_{\mathrm{rA}}^2}{p_{\mathrm{A}}}} \tag{22.9}$$

where p_{A} = the number of laboratories involved.

Analyzing the w ratios, several conclusions can be drawn: (1) In the CM 3 LopHC material, the pH measurements using the ESL method with filtering are in better agreement ($w \leq 1$) with respect to the pH measurements using the PFE method. This high trueness indicates that the routine ESL method with filtering is as accurate as the reference PFE method. (2) In "LopHC" materials with higher pH values (CM 4), the w ratios obtained are similar in both ESL methods. Although the w ratios calculated are not less than unity, they are still acceptable. However, it is evident that the ESL methods developed, mainly with filtering, are more accurate when they are used in materials that can be demonstrated as being LopHC (pH < =11)after the 90 +/− 5 day-curing period. This higher reliability of ESL methods in LopHC materials should be related with the different solid phases formed in these materials; for example, low-pH cementitious materials do not contain portlandite, but are mainly formed of C-S-H hydrates with a composition different from that observed in conventional cements [11–13]. Although the ESL methods give acceptable results in cementitious materials with more alkaline pore fluid pH values, some corrections considering the Ca content of the solution are possibly needed because the procedure followed in ESL method increases the Ca content in the solution, as suggested in [7, 10] due to the leaching effect of precipitated portlandite. A dilution effect of alkalies may also be involved. At last, by considering the higher pH value measured using the ESL method without filtering the suspension, the safety case scenario is guaranteed.

4 Conclusions

The main conclusions drawn from the results are:

1. The tests carried out have allowed validating the developed protocols in different low-pH cementitious materials.
2. The repeatability of the three pH measurement protocols evaluated, PFE, ESL without filtering and ESL with filtering, is very good since the resulting pH values are close or even lower to the error that can be assigned to the pH electrode.
3. The PFE technique is good enough to enable the acquisition of accurate pH values of solutions derived from low-pH cementitious materials and cementitious materials with high alkaline pore fluid pH values.
4. Although the PFE was not used in all the cementitious materials, the trueness obtained in both ESL procedures is good when compared with the pH measured from the PFE method (the reference method) in low-pH cementitious materials, and it is acceptable in cementitious materials with higher pore fluid pH values but some corrections considering the Ca content of the solution are possibly needed.

Acknowledgments Authors would like to thank the work carried out by the laboratories involved in this study: CBI (Sweden), CEA (France), CISDEM (Spain), CRIEPI (Japan), EMPA (Switzerland), JAEA (Japan), TFB (Switzerland), UJV (Czech Republic) and VTT (Finland).

References

1. Longuet P, Burglen L, Zelwer A (1973) The liquid phase of hydrated cement. Rev Matér Constr Trav Publics 676:35–41
2. Barneyback RS, Diamond S (1981) Expression and analysis of pore fluids from hardened cement pastes and mortars. Cement Concr Res 11:279–285
3. Yonezawa T, Ashworth V, Procter RPM (1988) Pore solution composition and chloride effects on the corrosion of steel in concrete. Corros Eng 44:489–499
4. Haque MN, Kayyali OA (1995) Free and water soluble chloride in concrete. Cement Concr Res 25:531–542
5. Thangavel K, Rengaswamy NS (1998) Relationship between chloride/hydroxide ratio and corrosion rate of steel in concrete. Cement Concr Compos 20:283–292
6. Pavlík V (2000) Water extraction of chloride, hydroxide and other ions from hardened cement pastes. Cement Concr Res 30:895–906
7. Castellote M, Alonso MC, Andrade C (2002) Determination of the OH- content in the aqueous phase of hardened cementitious materials by an empirical leaching method. Materiales de Construccion 52(265):39–56
8. Räsänen V, Penttala V (2004) The pH measurement of concrete and smoothing mortar using a concrete powder suspension. Cement Concr Res 34:813–820
9. Hidalgo A, García JL, Alonso MC, Fernández L, Andrade C (2005) Testing methodology for pH determination of cementitious materials. Application to low pH binders for use in HLNWR. R&D on low-pH cement for a geological repository. Workshop, Madrid, June 2005
10. Li L, Jingak N, Hartt WH (2005) Ex situ leaching measurement of concrete alkalinity. Cement Concr Res 35:277–283
11. Cau-dit-Coumes C, Courtois S, Nectoux D, Leclercq S, Bourbon X (2006) Formulating a low-alkalinity, high-resistance and low-heat concrete for radioactive waste repositories. Cement Concr Res 36:2152–2163
12. García Calvo JL, Hidalgo A, Alonso C, Fernández Luco F, (2010), Development of low-pH cementitious materials for HLRW repositories. Resistance against ground waters aggression, Cement Concrete Res 40:1290–1297
13. AlonsoMC, Garcia JL, Hidalgo A, Fernandez-Luco L (2010) Development and application of low-pH concretes for structural purposes in geological repository systems, Chap 10.3d. In: Geological repository systems, pp 286–322
14. ISO 5725-1993, Accuracy (trueness and precision) of measurement methods and results, International Standarization Organization, Geneva
15. Alonso MC, García Calvo JL, Pettersson S, Cuñado MA, Vuorio M, Weber H, Ueda H, Naito M, Walker C (2011) Development of an accurate methodology for measuring the pore fluid pH of low-pH cementitious materials. XXIII International congress on cement, Madrid

Index

F. Bart et al. (eds.), *Cement-Based Materials for Nuclear Waste Storage*,
DOI 10.1007/978-1-4614-3445-0, © Springer Science+Business Media New York 2013

Printed by Printforce, the Netherlands